Pulsed Laser Technologies - Advanced Techniques and Cutting-Edge Applications
http://dx.doi.org/10.5772/intechopen.1008084
Edited by Sulaiman Wadi Harun

Contributors
Abdalla M. Darwish, Abdellah Laazizi, Ahmad Haziq A. Rosol, Ayan Ghosh, Carlos Molpeceres, Carolina Teixeira, Cheng Xiau San, Darayas N. Patel, David Munoz-Martin, Dingyuan Tang, Heitor Teixeira, Hiyori Uehara, Hu Xiao, Kenji Goya, Lip Zhan Hong, Miguel Morales, Mouad Bensada, Peng Bi, Ramesh Chandra Das, Retna Apsari, Rongfei Wei, Ruben Pinheiro, Ryo Yasuhara, Seiji Shimizu, Sergey S. Sarkisov, Sergio Vela Liñán, Shigeki Tokita, Sulaiman Wadi Harun, Tan Sin Jin, Utpal Nundy, Wei Zhang, Wenhao Wei, Xiangling Tian, Yu Chen, Zian Cheak Tiu

First published in London, United Kingdom, 2026 by IntechOpen
IntechOpen is the global imprint of INTECHOPEN LIMITED, registered in England and Wales, registration number: 11086078, 167-169 Great Portland Street, London, W1W 5PF, United Kingdom

For EU product safety concerns: IN TECH d.o.o., Prolaz Marije Krucifikse Kozulić 3, 51000 Rijeka, Croatia, info@intechopen.com or visit our website at intechopen.com.

British Library Cataloguing-in-Publication Data
A catalogue record for this book is available from the British Library

Pulsed Laser Technologies - Advanced Techniques and Cutting-Edge Applications
Edited by Sulaiman Wadi Harun
p. cm.
Print ISBN 978-1-83635-070-5
Online ISBN 978-1-83635-069-9
eBook (PDF) ISBN 978-1-83635-071-2

Meet the editor

Sulaiman Wadi Harun obtained his B. E. in Electrical and Electronics System Engineering from Nagaoka University of Technology, Japan (1996), and his M. Sc. and Ph.D. in Photonics Technology from the University of Malaya (2001, 2004). With nearly two decades of research experience, he specializes in optical fiber devices, lasers, and sensors, exploring nanomaterials such as graphene, carbon nanotubes, and black phosphorus. Author of over 900 ISI-indexed papers with 8,000+ citations (h-index >50), he has supervised more than 80 Ph.D. graduates. A Fellow of the Academy of Sciences Malaysia, he is also an Academic Editor for the International Journal of Optics and an Adjunct Professor at Airlangga University. His awards include the Malaysian Rising Star Award (2016) and Optoelectronics Letters Best Paper Award (2023). Sulaiman Wadi Harun is included in the 2025 update of Stanford University's list of the Top 2% most-cited scientists in the world.

Contents

Preface

Pulsed lasers employ advanced techniques such as Q-switching, gain-switching, chirped pulse amplification (CPA), and mode-locking to produce high-energy, ultrashort light bursts with exceptional precision. These methods have enabled breakthrough applications in fields including material micro-processing, advanced imaging, high-capacity optical communications, biomedical surgery, and precision manufacturing. This book aims to highlight recent advances in laser technology, with a particular focus on pulse generation techniques and their state-of-the-art applications.

Optical fiber lasers have significantly influenced modern photonics, becoming indispensable tools across various industries and research fields. By exploiting the guiding properties of rare-earth-doped optical fibers, fiber lasers achieve efficient amplification and high beam quality through population inversion stimulated by diode-pumped architectures. Chapter 1 provides an overview of the physical foundations and key components of optical fiber pulsed lasers, presents the methods used to generate pulses, explores the major nonlinear effects at play, and concludes with a discussion of contemporary applications, future challenges, and directions.

The mid-infrared region encompasses two major laser systems: the CO_2 laser (9–11 μm) and the CO laser (5–7 μm). CO_2 lasers are extensively utilized in industrial applications such as cutting, welding, and surface processing due to their high efficiency and operational reliability. Although CO lasers require cryogenic cooling, they exhibit stronger absorption in structural materials and biological tissues, making them attractive for advanced material processing, biomedical applications, and environmental sensing.

Chapter 2 presents a theoretical investigation of the performance of a cryogenically cooled Q-switched CO laser employing a rotating-mirror Q-switching technique. By varying the mirror rotation frequency from 50 to 625 Hz, the laser's output power and spectral characteristics can be optimized, yielding a peak power of 200 kW at 500 Hz with a pulse width (FWHM) of 220 ns. The Q-switched pulse consists of approximately 100 CO laser lines spanning 5.0–6.2 μm, with 65 lines exceeding 100 W and a maximum of 25 kW. These features make the system highly suitable for nonlinear sum-frequency generation in crystals. This chapter provides new insights into the operational physics of Q-switched CO lasers and their potential for advanced spectroscopic applications.

Recent advancements in low-dimensional nanomaterials have greatly broadened the scope of saturable absorbers (SAs) in pulsed fiber laser systems, facilitating the efficient generation of ultrashort optical pulses via passive Q-switching and mode-locking techniques. Chapter 3 presents a comprehensive overview of the fundamental physical mechanisms—such as saturable absorption, the Kerr effect, and nonlinear optical interactions—and outlines key characterization techniques, including the

Z-scan and double-balanced detection methods. Special emphasis is placed on emerging classes of functional materials, including zero-dimensional quantum dots, one-dimensional nanowires, and two-dimensional materials such as graphene, transition-metal dichalcogenides, black phosphorus, topological insulators, and perovskites. The chapter also discusses their synthesis approaches, distinctive optical properties, and strategies for seamless integration into fiber laser systems.

Fiber lasers operating in the mid-infrared region have garnered considerable attention for applications in biomedical diagnostics, optical communications, and remote sensing. Chapter 4 demonstrates Q-switched laser operation near the 2 μm wavelength region using niobium carbide–based (Nb_2CT_x) MXene thin films as SAs. In this study, Nb_2CT_x MXene was integrated into two different host polymers—polyvinyl alcohol (PVA) and polyethylene oxide (PEO)—to form composite SA films. The Q-switched laser achieved pulse widths of 1.995 μs and 1.942 μs for Nb_2CT_x–PEO and Nb_2CT_x–PVA, respectively, both markedly shorter than the 13.3–4.4 μs range reported in previous studies. These results underscore the strong potential of MXene-based SAs for efficient pulsed laser generation in the 2 μm region.

Chapter 5 investigates the single-mode–multimode–single-mode (SMS) fiber structure as a simple, compact, and cost-effective approach for realizing Q-switched and mode-locked fiber lasers. Acting as an artificial SA through nonlinear multimode interference and the Kerr effect, the SMS configuration enables ultrashort pulse generation when integrated into a ring erbium-doped fiber laser (EDFL) and optimized using a polarization controller. Experiments in both anomalous and normal dispersion regimes demonstrated Q-switched operation with step-index multimode fibers and mode-locked operation with graded-index fibers, achieving conventional and bound soliton regimes. The SMS-based absorber also exhibited a higher damage threshold compared with conventional thin-film SAs.

Chapter 6 examines spectral sideband formation in soliton fiber lasers, a key phenomenon driven by multiple mechanisms. Kelly sidebands result from interference between solitons and dispersive waves, while modulation instability (MI) generates sidebands in both dispersion regimes. Additional sidebands can emerge from periodic variations in soliton parameters or polarization instability. Understanding these effects is vital for analyzing soliton dynamics and improving the performance and stability of ultrafast fiber lasers for telecommunications and high-power applications.

Chapter 7 reviews recent progress in mid-infrared (MIR) laser technology for materials processing, focusing on fiber Bragg gratings (FBGs) fabricated using femtosecond laser writing. The plane-by-plane technique enables precise, stable FBGs to be directly formed in fluoride glass fibers, forming compact, wavelength-selective cavities. These advances have produced efficient MIR fiber lasers operating near 2.8–2.92 μm, ideal for cutting, surface modification, and fusion splicing with minimal thermal damage. MIR lasers offer superior absorption and processing capability for polymers and glasses, making them a powerful tool for high-precision photonic manufacturing across diverse materials.

Chapter 8 introduces the Teixeira-Pinheiro Technique (2025), a novel approach using picosecond Nd:YAG lasers to remodel post-surgical hypertrophic scars, particularly in patients after cleft-lip correction. Based on a prospective study of over thirty patients, the treatment produced significant clinical and aesthetic improvements with minimal side effects. The technique leverages controlled photomechanical effects suitable for delicate facial tissue, establishing picosecond Nd:YAG laser therapy as a first-line option for managing surgical and post-traumatic scars or contractures.

Chapter 9 presents a Multiphysics modeling approach for Nd:YAG pulsed laser welding of X30Cr13 stainless steel. The study simulates temperature fields, weld-pool geometry, and residual stresses using heat-transfer equations with realistic boundary conditions and material properties. Finite element simulations are validated against experimental data, revealing detailed insights into temperature distribution, melt pool behavior, and stress formation. The model serves as a predictive tool for optimizing welding parameters and minimizing defects, enhancing the reliability and performance of welded components in industrial applications.

Chapter 10 presents theoretical and numerical models for ultrashort-pulse laser ablation of metals. Because pulse durations are shorter than the lattice relaxation time, a two-temperature model is used to describe electron and lattice temperature dynamics separately. Material removal occurs via spallation near the threshold fluence and by phase explosion at higher fluences. Model predictions, validated against experimental data for aluminum, show strong agreement, offering a deeper understanding of ultrashort-pulse ablation mechanisms and their dependence on laser fluence.

Chapter 11 explores the use of Q-switched lasers for concurrent multibeam multi-target pulsed laser deposition of high-entropy alloy (HEA) films. A six-beam setup incorporating magnets and substrate heating enhanced plasma confinement and deposition efficiency. Films containing up to six metals (Au, Ag, Cu, Al, Ti, W) were grown on various substrates, forming crystalline nanoparticles and alloy mixtures. Heating to 300°C promoted greater alloy diversity. The resulting HEA films enhanced the photoluminescence of $CsPbBr_3$ perovskite quantum dots by 6–16 times through plasmonic resonance and resonant coupling effects, demonstrating strong potential for optical sensing applications.

Sulaiman Wadi Harun
Department of Electrical Engineering,
University of Malaya,
Kuala Lumpur, Malaysia

Chapter 1

Principles of Optical Fiber Pulsed Lasers and Their Advanced Applications

Wei Zhang

Abstract

Optical fiber lasers have significantly influenced modern photonics, becoming indispensable tools across various industries and research fields. By exploiting the guiding properties of rare-earth-doped optical fibers, fiber lasers achieve efficient amplification and high beam quality through population inversion stimulated by diode-pumped architectures. Pulsed fiber lasers, in particular, have evolved to offer precise temporal control, high peak powers, and ultrashort pulse durations ranging from nanoseconds to femtoseconds. Various pulse generation mechanisms, including gain-switching, active and passive Q-switching, and mode-locking, are discussed, highlighting their respective operating principles, pulse characteristics, and application scenarios. Critical system components—such as pump sources (typically diode lasers), fiber resonators utilizing Bragg gratings or ring resonators, modulators, saturable absorbers, and effective dispersion management—are examined in detail, emphasizing their roles in enhancing performance, reliability, and efficiency. Nonlinear optical phenomena inherent to fiber lasers, including self-phase modulation, stimulated Raman scattering, and four-wave mixing, significantly influence pulse formation and stability, necessitating careful dispersion and nonlinear management strategies. Advanced methods such as chirped pulse amplification (CPA) have facilitated the generation of high-energy femtosecond pulses crucial for demanding scientific and industrial applications. Pulsed fiber lasers have rapidly penetrated diverse sectors including precision materials processing, biomedical surgery and imaging, high-capacity optical communications, LiDAR-based remote sensing, and fundamental scientific research. The chapter concludes with a forward-looking perspective, emphasizing the ongoing challenges and anticipated technological innovations that will continue driving fiber laser development and expanding their capabilities and application domains.

Keywords: pulsed fiber lasers, Q-switching, mode-locking, chirped pulse amplification, nonlinear optics, ultrafast lasers

1. Introduction

The development of optical fiber lasers has had a profound impact on modern photonics and its various applications. Fiber lasers leverage the light-guiding properties of optical fibers that are doped with rare-earth ions to achieve efficient amplification. Over time, fiber laser technology has evolved from continuous-wave sources to sophisticated pulsed systems. Pulsed fiber lasers are now essential tools in many areas that require precise temporal control, high peak power, or ultrashort pulse durations. They are employed in research laboratories, medical facilities, telecommunications infrastructures, and manufacturing lines. One of the key reasons for the popularity of fiber lasers is their waveguide geometry, which naturally provides a stable single-mode output with high beam quality. The fiber geometry also offers an excellent ability to dissipate heat along its length. This is especially advantageous under high-power or high-energy operation [1], where other laser configurations might face severe thermal lensing or mechanical alignment issues. Fiber lasers can be pumped by low-cost, reliable, high-power laser diodes, typically launched into the fiber core or cladding, and the overall footprint of fiber-based systems is generally smaller than bulk solid-state lasers. The confinement of light within a narrow fiber core also ensures robust alignment once the fiber is in place. These inherent advantages have spurred the development of pulsed fiber lasers that can deliver pulses ranging from nanoseconds to femtoseconds.

Pulsed fiber lasers are achieved by controlling the optical gain or cavity losses in time, or by forcing the phase of different longitudinal modes to synchronize. There are multiple ways to generate pulses in fiber lasers, such as gain-switching, Q-switching, and mode-locking. Each mechanism has distinct operating principles, pulse durations, peak powers, and repetition rate ranges. For instance, gain-switching uses the dynamic of the pump diode or pump modulation, Q-switching works by modulating cavity losses, and mode-locking relies on phase-locking multiple modes inside the resonator to produce very short pulses in the picosecond or femtosecond regime. The rapidly expanding market for pulsed fiber lasers is driven by their use in precise material processing, high-resolution spectroscopy, advanced diagnostics, medical imaging, communications, and laser-based sensing. As requirements for higher average powers and shorter pulse durations continue to rise, fiber technology has adapted to meet these needs with various large-mode-area fiber designs, specialized dopants, chirped pulse amplification strategies, and effective thermal management. This chapter provides an overview of the physical foundations and key components of optical fiber pulsed lasers, presents the methods used to generate pulses, explores the major nonlinear effects at play, and concludes with a discussion of contemporary applications as well as future challenges and directions.

2. Fundamentals of optical fiber lasers

Optical fiber lasers rely on rare-earth-doped glass fibers as their gain medium. Traditionally, silica fibers doped with ions like erbium, ytterbium, or thulium are employed to produce laser emission in wavelength ranges relevant to telecommunications, materials processing, or medical diagnostics. The principle of operation is based on population inversion, in which dopant ions in the fiber core are excited to higher energy states using pump light. The pump light is commonly supplied by semiconductor laser diodes, either directly into the core for single-mode fibers or into an inner cladding that surrounds the core for higher power double-clad fibers. Once the inversion is achieved, stimulated emission at a specific transition wavelength leads to amplification and, in the presence of resonator feedback, laser oscillation.

Silica-based fibers offer strong mechanical stability and low propagation loss. By confining the optical mode inside a flexible waveguide, they simplify alignment and reduce sensitivity to environmental vibrations. Another practical advantage of fiber lasers lies in their ability to dissipate heat along a long length of fiber, rather than concentrating it in a small region of a crystal. This thermal management leads to improved beam quality under high power. Despite these benefits, care must be taken with nonlinear effects that arise at higher intensities over long interaction lengths. Phenomena such as stimulated Raman scattering, stimulated Brillouin scattering, and self-phase modulation can either be detrimental or harnessed for specific applications, depending on the design.

In the simplest picture, a fiber laser is pumped by a pump beam that injects photons at a wavelength well-absorbed by the dopant ions. These ions, after absorbing the pump photons, are elevated to higher energy levels. Through radiative and non-radiative processes, the system reaches a steady state where there is a net gain for photons at the laser transition wavelength. A feedback mechanism provided by mirrors or fiber Bragg gratings (FBG) completes the laser cavity. The generated laser output can be continuous-wave or pulsed, depending on how the gain or loss is modulated in time. The power scaling of fiber lasers has been achieved in part by double-clad fiber technology [2], in which the pump light is guided in a larger inner cladding region, while the signal remains confined in a smaller doped core. This arrangement increases the effective pump absorption and allows the use of multiple pump diodes. Another route to higher power is the use of large-mode-area fibers, which increase the core diameter while maintaining single-mode guidance. These fibers reduce nonlinear interactions by lowering the intensity and decreasing the effective interaction length. As a fundamental background, the doping concentration, core geometry, and fiber length must be carefully chosen to match the desired pump absorption, gain bandwidth, and operational mode area to yield stable laser output.

3. Components of a fiber pulsed laser system

The efficient operation of a pulsed fiber laser relies on multiple carefully selected components. These include the pump source, the design of the fiber resonator, the modulators or saturable absorbers used for pulse formation, and the management of dispersion or nonlinearity along the length of the fiber.

3.1 Pump sources

Fiber lasers are typically pumped by laser diodes. These diodes can be operated at wavelengths that match the absorption peak of the doped fiber core. Single-mode diodes can directly couple light into the fiber core for lower power systems. For higher power, cladding-pumped designs are favored. In a cladding-pumped configuration, multiple pump diodes feed light into a larger inner cladding, while the signal propagates in a smaller doped core. Alternatively, Raman pumping is sometimes employed to shift the pump wavelength to more favorable absorption bands within the fiber. The choice of pump wavelength and architecture is influenced by the dopant (for example, ytterbium- or erbium-doped fiber), the desired output power, and the available pump diode technology.

Figure 1 shows a double-clad fiber laser in an end-pumped configuration, a widely used architecture for high-power fiber laser systems. The structure consists

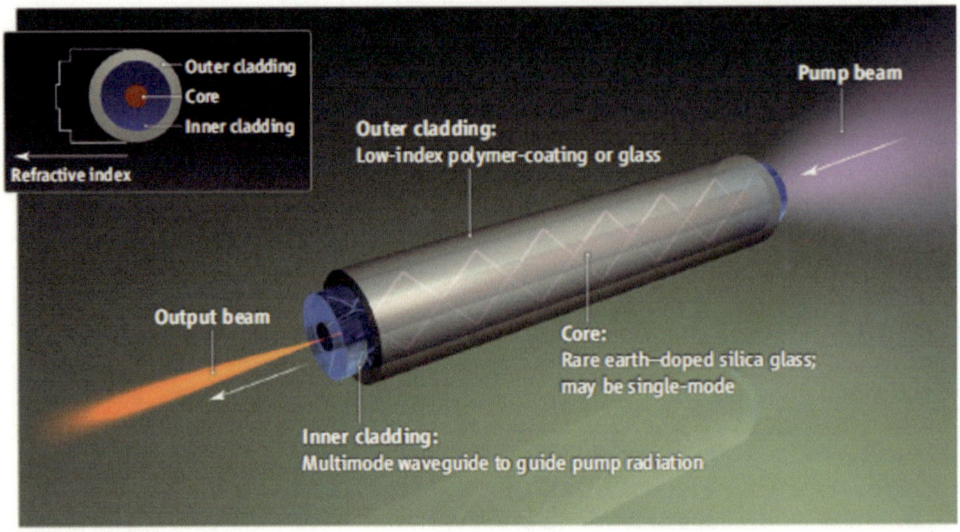

Figure 1.
Schematic of a double-clad fiber laser in an end-pumped configuration. [Reprinted/Adapted] with permission from [2].

of three main layers: the core, the inner cladding, and the outer cladding. The core, made of rare-earth-doped silica glass, serves as the gain medium where stimulated emission occurs, enabling laser amplification. It can be single-mode, ensuring high beam quality. Surrounding the core is the inner cladding, which functions as a multimode waveguide to guide pump radiation from the pump beam into the core. This allows for efficient absorption of pump energy by the doped ions in the core. The outermost layer, the outer cladding, is made of a low-index polymer coating or glass, ensuring total internal reflection (TIR) within the inner cladding and maximizing pump absorption efficiency. In this end-pumped configuration, the pump beam is injected at one end of the fiber, propagating through the inner cladding and gradually transferring energy to the core. The amplified output beam emerges from the other end, typically featuring high brightness and efficiency. This double-clad fiber design allows for high pump absorption, excellent heat dissipation, and power scaling, making it ideal for industrial, medical, and defense applications, including materials processing, laser surgery, and directed energy systems.

3.2 Fiber resonators

In fiber lasers, the resonator is formed by elements such as FBG or partially reflective mirrors spliced at the fiber ends. A simple linear resonator can use a high reflector grating at one end and a partial reflector at the other, allowing the output beam to exit. Alternatively, a ring resonator can be used for certain designs [3–5]. The fiber waveguide ensures that alignment of the cavity does not drift over time. Such stability is one of the hallmarks of fiber lasers. Resonator design also involves dispersion engineering and doping profiles that match the cavity length, pump power, and intended repetition rate. The overall performance of the laser, including its threshold, output power, beam quality, and pulse characteristics, depends on the resonator's arrangement.

DOI: http://dx.doi.org/10.5772/intechopen.1010288

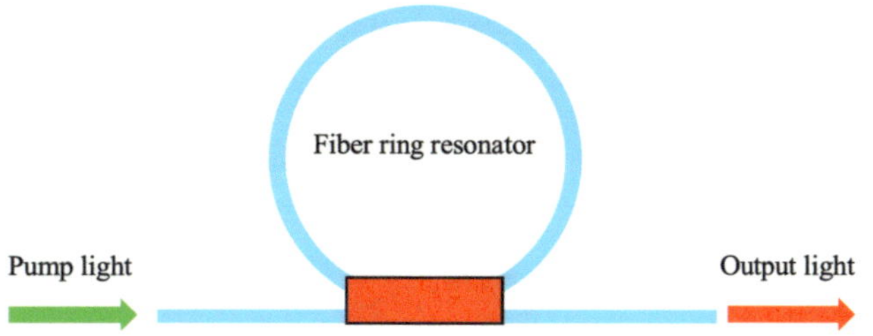

Figure 2.
Schematic of a ring resonator fiber laser.

Figure 2 shows a fiber ring resonator (FRR) system, a fundamental optical structure used for filtering, sensing, and nonlinear optical applications. The system consists of an optical fiber loop that acts as a resonant cavity, coupled to a straight waveguide through an interaction region (shown in red). The pump light (green arrow) enters the waveguide and interacts with the resonator, where certain resonant wavelengths are selectively enhanced due to constructive interference. The resonator enhances specific optical frequencies while suppressing others, leading to a modulated output light (red arrow). This configuration enables precise wavelength filtering, high-quality factor (Q-factor) resonance effects, and optical signal processing. Fiber ring resonators are widely used in applications such as laser stabilization, optical switching, wavelength division multiplexing (WDM), and nonlinear optics. The coupling efficiency, ring size, and material properties determine the spectral characteristics and performance of the system.

3.3 Modulators and saturable absorbers

Modulators are indispensable for active Q-switching or active mode-locking. An electro-optic modulator employs the Pockels effect in certain crystals, and by applying a voltage, one can rapidly change the loss or phase inside the fiber cavity. An acousto-optic modulator diffracts the laser beam in response to an acoustic wave, introducing time-varying loss. In passive mode-locking or passive Q-switching, saturable absorbers provide an intensity-dependent transmission [6]. Materials such as semiconductor saturable absorber mirrors (SESAMs) [7] or nanomaterial-based saturable absorbers (including graphene or carbon nanotubes) [8] have emerged in fiber laser designs to produce stable short pulses without the need for external driving electronics.

3.4 Dispersion management and nonlinear effects

Dispersion in an optical fiber can be normal (positive group-velocity dispersion) or anomalous (negative group-velocity dispersion), depending on both wavelength and fiber design. Precisely managing dispersion is central to controlling pulse evolution, particularly in mode-locked lasers. Under high-intensity pulsed operation, nonlinear effects—including self-phase modulation (SPM), cross-phase modulation (XPM), stimulated Raman scattering (SRS), and four-wave mixing (FWM)—arise from the small fiber core diameter and long propagation distance. These phenomena can broaden a pulse's spectrum, shift its wavelength, and influence pulse stability. The generalized nonlinear Schrödinger equation (GNLS) has been used to model pulse propagation in the optical fiber [9]:

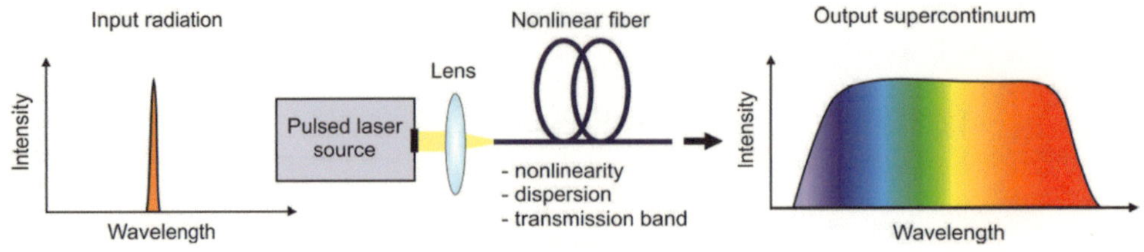

Figure 3.
Schematic setup for supercontinuum generation. [Reprinted/Adapted] with permission from [10].

$$\begin{aligned}\frac{\partial A(z,t)}{\partial z}+\frac{\alpha}{2}A-iIFT\left\{\left[\beta(\omega_0+\Omega)-\beta(\omega_0)-\Omega\beta_1(\omega_0)\right]\tilde{A}(z,\Omega)\right\}\\ =i\gamma\left(1+\frac{i}{\omega_0}\frac{\partial}{\partial T}\right)\left[A(z,t)\int_{-\infty}^{t}R(t-t')\left|A(z,t')\right|^2 dt'\right]\end{aligned} \quad (1)$$

where A (z, t) is the electric field envelope as a function of distance along the fiber z and retarded time t, α is the fiber loss, $\beta(\omega)$ is the wavenumber as a function of frequency, and $\beta_1(\omega) = d\beta/d\omega$ is the inverse group velocity. We use IFT to denote the inverse Fourier transform, where Ω is the transform variable and the tilda indicates the Fourier transform. The third term on the left-hand side of Eq. (1) accounts for dispersion, and the right-hand side accounts for the nonlinear contribution.

A balanced combination of dispersion and nonlinearity enables the formation of solitons or the generation of supercontinuum light. In soliton lasers, SPM counteracts dispersion, stabilizing ultrashort pulses. For supercontinuum sources, multiple nonlinear processes transform an initially narrower pulse into a broad-spectrum spanning hundreds of nanometers or more, which is advantageous in spectroscopy, optical coherence tomography, and other metrology applications. However, the same mechanisms can limit performance if left unchecked. SRS can become parasitic unless carefully managed, and stimulated Brillouin scattering (SBS) often restricts output power in narrow-linewidth lasers. Designing a fiber system that harnesses desired nonlinear processes while suppressing detrimental ones is thus key to achieving high efficiency, stability, and reliable operation in pulsed fiber lasers.

Figure 3 shows the process of supercontinuum generation using a pulsed laser source and a nonlinear optical fiber. Initially, a narrowband input radiation with a high peak intensity is emitted by the pulsed laser source. This laser pulse is then focused using a lens into a nonlinear fiber, where several key physical effects—such as optical nonlinearity, dispersion, and the fiber's transmission properties—contribute to the spectral broadening of the input light. As the pulse propagates through the fiber, nonlinear effects interact to significantly expand the spectral bandwidth of the initial pulse. The resulting output is a supercontinuum, characterized by a broad, continuous spectrum that spans multiple wavelengths, as illustrated by the rightmost graph.

4. Pulse generation mechanisms

Pulses in fiber lasers are generated by actively or passively modulating gain, loss, or mode phases within the resonator. Four major techniques are typically employed.

4.1 Gain-switching

Gain-switching is one of the simplest methods for generating laser pulses. In this approach, the gain medium is driven by a pump source that is turned on and off at high speeds. A common strategy is to modulate the current of a laser diode pump, causing the gain to rise rapidly when the current is switched on. Once the intracavity photon density reaches a certain threshold, a short burst of stimulated emission depletes the inversion, resulting in a pulse. This mechanism can be implemented using direct diode modulation for fiber lasers, making it straightforward and cost-effective. Gain-switching can produce pulses on the order of tens of nanoseconds down to subnanosecond durations, but controlling the precise temporal shape and repetition rate can be more challenging compared to other pulsing methods. The pump transition must be carefully matched so that the gain medium quickly accumulates population inversion and then releases it in the form of a pulse. While not as flexible for high-energy or ultrashort pulses, gain-switched fiber lasers do find applications in low-to-moderate power tasks, or in seed sources for further amplification.

Figure 4 shows a fiber-based laser system utilizing a rare-earth-doped fiber as the gain medium. The system is driven by a pulsed pump source, which provides the necessary energy to excite the rare-earth ions within the fiber, and the pump is applied in bursts rather than continuously. This is a key requirement for gain-switching. An optical isolator is placed immediately after the pump source to prevent unwanted back reflections, which could destabilize the laser operation. Two fiber Bragg gratings (FBG1 and FBG2) are strategically positioned within the system to act as wavelength-selective reflectors, forming a distributed feedback mechanism that helps in defining the laser's output wavelength and improving efficiency.

Table 1 summarizes the output characteristics of gain-switched fiber lasers, including fiber types, pump and laser wavelengths, pulse energy, slope efficiency, pulse widths, and repetition rates. Tm-doped silica achieves the highest pulse energy (14.7 mJ), while Nd-doped silica operates at the highest repetition rate (2500 kHz). The data highlights the versatility of gain-switched fiber lasers across different wavelengths, making them ideal for moderate-power pulsed applications or as seed sources for further amplification.

4.2 Q-switching

Q-switching relies on modulating the intracavity losses to store energy in the gain medium before suddenly releasing it. In a low-Q state, the cavity quality factor is deliberately reduced by introducing high losses with a modulator or saturable absorber. The inversion level (and hence stored energy) continues to

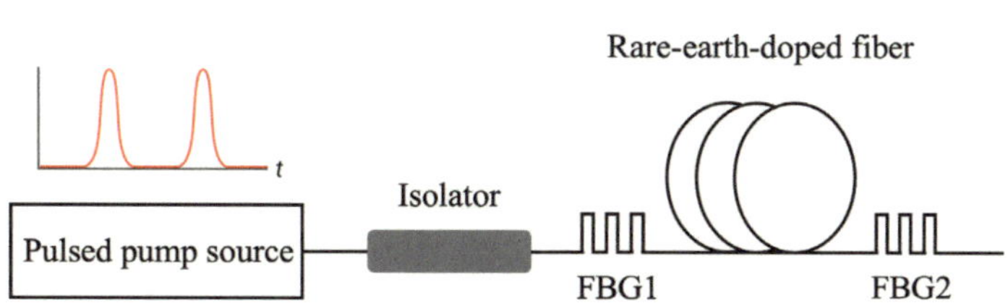

Figure 4.
Basic setup of an all-fiber gain-switched fiber laser.

Fiber type	Pump λ (nm)	Laser λ (nm)	Pulse energy (mJ)	Slope efficiency (%)	Pulse widths (ns)	Repetition rates (kHz)
Nd-doped silica	500	1060	1.84e-6	32 (57)	40	2500
Tm-doped silica	1064	1840–1940, 2000, 2040	14.7	40 (71)	330	—
Tm-doped silica	790	1909–2018	10.1	30 (74)	300	—
Er-doped ZBLAN	791	2700–2770	1.9	13.5 (47)	200	—
Er/Yb doped silica	900	1040–1046, 1535–1541	1.38	15 (17)	—	—
Tm-doped silica	1550	1990, 2000, 2044	0.035	50 (65)	10	500
Er-doped fluoride	974+ 1550	845	0.002	—	2440	200
Yb-doped silica	965	1080	0.036	36 (40)	770	50
Yb-doped silica	915	1064	0.15	60 (70)	200	210
Ho-doped silica	1909	2106	0.003	44 (49)	150	80
Tm-doped silica	1914	1940, 2020	10	80 (81)	61	50
Ho-doped silica	1950	2104, 2112	0.016	65 (70)	38	600
Ho-doped fluoride	1150	2074	0.006	4.8 (9)	460	30
Tm-doped silica	790+ 1053	2018	2	29.1 (74)	390	100

Table 1.
Reported output features of the gain-witched fiber lasers. [Reprinted/Adapted] with permission from [11].

build because the cavity is not allowing significant laser emission. At a specified moment, the losses are switched to very low, creating a high-Q state that allows the stored energy to be emitted in a short, intense pulse. Q-switched fiber lasers can be either actively Q-switched or passively Q-switched. Active Q-switching often uses electro-optic or acousto-optic modulators, whereas passive Q-switching uses saturable absorbers that change transmission once the intensity is high enough to bleach them. Q-switched fiber lasers typically produce pulses in the nanosecond to microsecond range at repetition rates ranging from a few hertz to megahertz, with energies from microjoules to millijoules depending on design. They are widely used in industrial settings for marking, engraving, and other micromachining processes. The ease of obtaining high peak powers and relatively short pulses makes Q-switched fiber lasers attractive for numerous applications that do not demand ultrafast timescales.

4.2.1 Active Q-switching

Active Q-switching usually uses an electro-optic or acousto-optic modulator as shown in **Figure 5**. In an electro-optic Q-switch, the application of a high voltage across a Pockels cell changes the polarization state of light and thereby switches the cavity losses. When the cavity is set to a low-Q state, the inversion accumulates. By quickly switching the voltage to reduce the loss, the stored energy is released in a single, intense pulse. Acousto-optic Q-switches diffract light by generating acoustic waves in a crystal, and switching the acoustic wave on or off modulates the cavity losses. The achievable pulse energy depends on the doping level, fiber length, and pump power, as well as on the dynamic range of the modulator.

Figure 6 shows a chronological overview of advancements in all-fiber Q-switching technologies and intracavity Q-factor modulators from 1986 to the present. The timeline is divided into two main categories: the development of all-fiber Q-switching technologies (highlighted in blue) and the evolution of intracavity Q-factor modulators (highlighted in orange). In the early stages, from 1986 to 1995, innovations such as evanescent field-based modulators and fiber interference

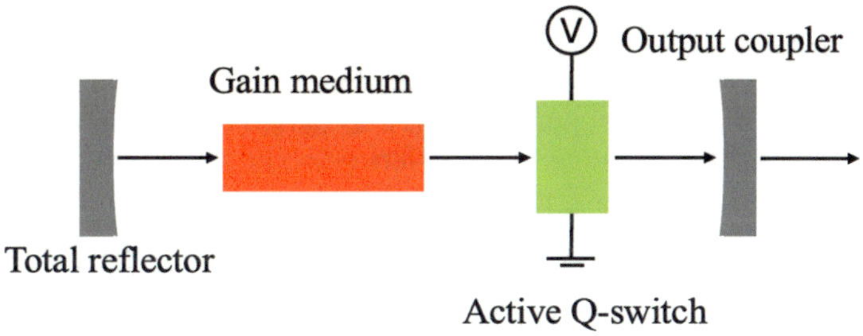

Figure 5.
Electro-optic active Q-switch lasers.

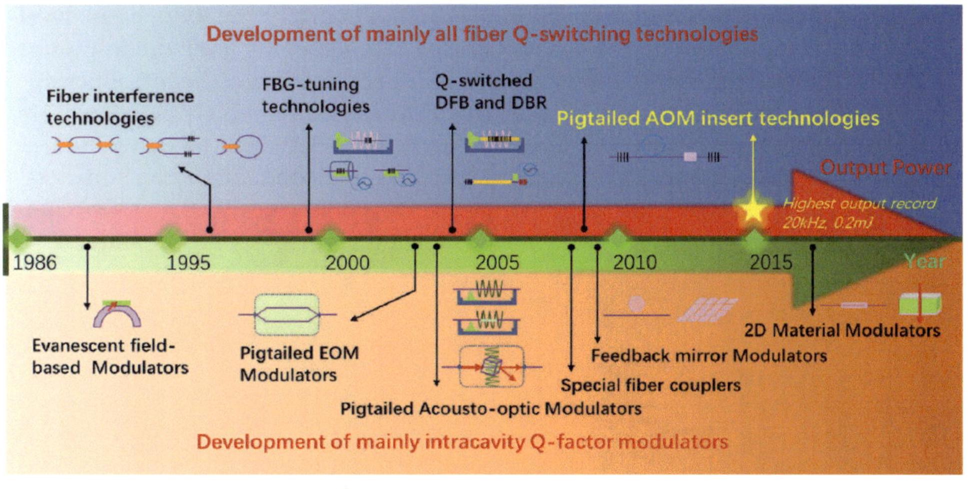

Figure 6.
Development tree of all-fiber actively Q-switched laser technologies and Q factor optic modulators. [Reprinted/Adapted] with permission from [12].

technologies were introduced. By the late 1990s and early 2000s, pigtail electro-optic modulators (EOM) and FBG tuning technologies were developed, enabling improved wavelength control and stability. Around 2005, the integration of Q-switched distributed feedback (DFB) and distributed Bragg reflector (DBR) lasers, along with pigtail acousto-optic modulators (AOM), played a significant role in advancing fiber laser performance. The next phase, from 2010 onwards, saw the introduction of special fiber couplers, feedback mirror modulators, and emerging 2D material-based modulators, marking a shift toward compact, efficient, and high-power fiber lasers. The figure also highlights a significant milestone—achieving the highest output power record of 20 kHz and 0.2 mJ—demonstrating the progress made in actively Q-switched fiber laser technology. This development timeline showcases the continuous innovation in fiber-based Q-switching, paving the way for applications in high-power pulsed laser systems used in industrial, medical, and scientific fields.

4.2.2 Passive Q-switching

Passive Q-switching is achieved by incorporating a saturable absorber into the laser cavity, as shown in **Figure 7**. In the low-intensity regime, the absorber imposes large losses, preventing the laser from reaching threshold. Once the population inversion grows sufficiently high, the stored energy in the gain medium causes an intracavity intensity surge, bleaching the absorber. A large fraction of the stored energy is then emitted as a short pulse. Passive Q-switching is typically simpler and less expensive than active methods, but it offers less control over repetition rate or pulse duration. Material choices for the saturable absorber can be crystals doped with ions such as Cr^{4+} in YAG or novel two-dimensional materials such as graphene. Energy and peak power scaling in Q-switched fiber lasers rely on large-mode-area fiber configurations, effective thermal management, and controlling nonlinearities that might become problematic at higher pulse intensities.

Figure 8 shows a timeline of major advancements in passive Q-switching for Er:Yb:glass lasers, highlighting key breakthroughs from 1995 to the present. The timeline is segmented into different periods, showcasing the progressive evolution of Q-switching technologies and material innovations. From 1995 to 2000, early developments included the first demonstration of Co:LMA achieving laser-diode-pumped, single-longitudinal-mode operation in 2000. This period also saw the introduction of SESAM (Semiconductor Saturable Absorber Mirror) and double-pass resonators, setting the foundation for efficient passive Q-switching. Between 2001 and 2005, significant milestones were reached, such as the PbS semiconductor quantum-dot-doped glass (2001) and Co:ASL single-longitudinal-mode laser (2003). Q-switching efficiency was greatly improved, with YAG achieving up to 37% (2004), and new

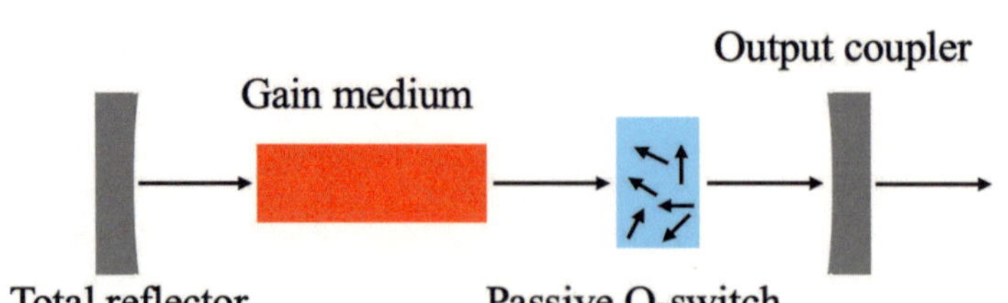

Figure 7.
Passive Q-switch diagram lasers.

DOI: http://dx.doi.org/10.5772/intechopen.1010288

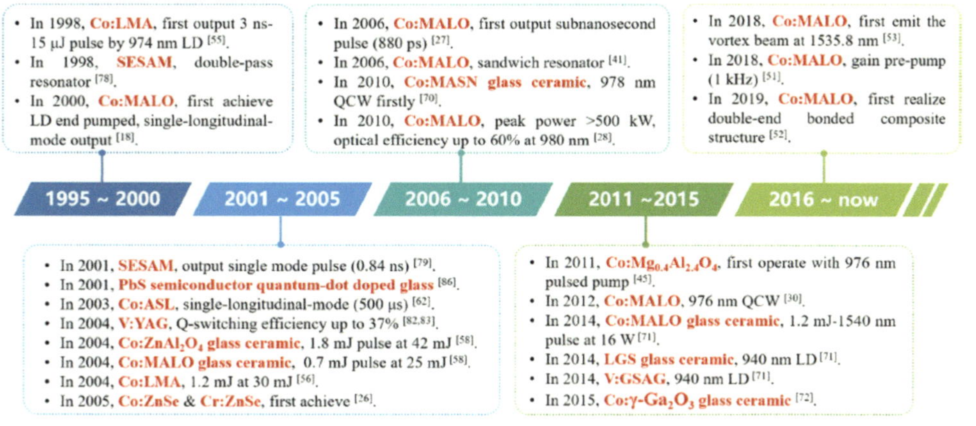

Figure 8.
Timeline of the major advances for passive Q-switches in Er:Yb:glass lasers in recent year. [Reprinted/Adapted] with permission from [13].

absorber materials like $Co:ZnAl_2O_4$ glass ceramic were introduced, demonstrating impressive output characteristics. The 2006–2010 period marked advancements in ultrashort pulse generation, with Co:MALO producing subnanosecond pulses (2006) and Co:MASN glass ceramic operating at 978 nm QCW (2010). Additionally, the peak power of Co:MALO lasers exceeded 500 kW, achieving high optical efficiency. During 2011–2015, new materials such as $Co:MgAl_2O_4$ (2011) and LGS glass ceramic (2014) were developed, offering enhanced performance. Notably, Co:MALO demonstrated QCW operation at 976 nm (2012) and a 1.2 μm output at 14 W (2014). From 2016 to the present, cutting-edge innovations have emerged, including Co:MALO emitting vortex beams at 1535.8 nm (2018) and achieving gain pre-pumping at 1 kHz (2018). In 2019, Co:MALO introduced a double-end bonded composite structure, demonstrating the latest evolution in passive Q-switching technologies. Overall, this timeline showcases the steady progression in passive Q-switching materials and laser performance, paving the way for advanced applications in high-power pulsed laser systems, telecommunications, and precision laser machining.

Table 2 summarizes the characteristics of passively Q-switched crystals used in Er:Yb:glass fiber lasers, listing their chemical formulas, excited-state lifetimes, absorption bands, and thermal conductivities. Co:ZnSe and Cr:ZnSe exhibit the longest excited-state lifetimes, making them suitable for mid-infrared applications, while Co:MALO has the highest thermal conductivity (25 W/mK), beneficial for heat dissipation in high-power lasers. The choice of saturable absorber material significantly impacts pulse energy, wavelength range, and thermal management, making it essential for optimizing passive Q-switching performance in fiber laser systems.

4.3 Mode-locking

Mode-locking is generally the technique of choice for generating ultrashort pulses, often with durations on the order of femtoseconds to picoseconds. A typical mode-locked fiber laser repetitively produces pulses at a rate determined by the cavity length and the group velocity of light in the fiber. By enforcing a fixed phase relationship among many longitudinal modes of the cavity, the output light interferes constructively in time at regular intervals, creating a train of pulses. In active

Crystals	Chemical formulas	Excited-state lifetimes (ns)	Absorption bands (μm)	Thermal conductivities (W/(mK))
Co:MALO	$Co^{2+}:MgAl_2O_4$	350	1.1–1.7	25
Co:LMA	$Co^{2+}:LaMgAl_{11}O_{19}$	210	1.0–1.6	5.4/3.1 (a/c)
Co:ASL	$Co_{3+}:La_xSr_{1-x}Mg_{x-y}Co_yAl_{12-x}O_{19}$	210	1.1–1.6	6.3/4.4 (a/c)
Co:ZnSe	$Co^{2+}:ZnSe$	2.9e5	1.3–2.0	16
Cr:ZnSe	$Cr^{2+}:ZnSe$	8e3	1.4–2.2	14
V:YAG	$V^{3+}:Y_3Al_5O_{12}$	20 ~ 30	1.0–1.45	11/2 (a/c)

Table 2.
Characteristics of passively Q-switched crystals commonly used in Er:Yb:glass fibers. [Reprinted/Adapted] with permission from [13].

mode-locking, an external modulator (often driven by an RF signal) imposes a periodic modulation of loss or phase within the cavity that synchronizes the modes. In passive mode-locking, a saturable absorber or nonlinear mechanism is used to automatically shape and shorten pulses. The saturable absorber discriminates between low-intensity light (which experiences high losses) and high-intensity light (which encounters lower losses), thus favoring short spikes in intensity that eventually form stable pulses. Passive mode-locking can lead to pulse durations ranging from a few picoseconds down to tens of femtoseconds, depending on dispersion management, nonlinearity, and gain bandwidth. Fiber lasers are well-suited to mode-locking because the waveguiding structure supports consistent overlap between the saturable absorber and the mode. Numerous techniques, including nonlinear polarization evolution or semiconductor saturable absorber mirrors, are employed. Mode-locked fiber lasers can serve as compact sources of ultrashort pulses for spectroscopy, frequency metrology, biomedical imaging, and fundamental physics studies.

Table 3 summarizes the characteristics of mode-locked Er^{3+}-, Ho^{3+}/Pr^{3+}-, and Dy^{3+}-doped fluoride fiber lasers operating in the 2.8–3.5 μm spectral region, listing mode-locking techniques, pulse durations, wavelengths, output power, and peak power. Passive mode-locking methods, such as SESAM, nonlinear polarization rotation (NPR), and semiconductor saturable absorbers (SA), enable ultrashort pulse generation with durations down to sub-picosecond levels, while techniques like BP and Bi_2Te_3 absorbers achieve high peak powers. These mode-locked fiber lasers serve as compact sources of ultrashort pulses, making them valuable for applications in mid-infrared spectroscopy, biomedical imaging, and ultrafast optical research.

4.3.1 Active mode-locking

Active mode-locking incorporates an external modulation drive, often at a radio frequency that matches the cavity round trip. Through electro-optic or acousto-optic devices, a periodic loss or phase modulation is applied. Each time the pulse circulates in the cavity, the modulation gate retimes and reshapes it. While active mode-locking can be simpler for synchronization with external systems, it is less likely to produce the extremely short durations achievable with passive mode-locking because the bandwidth can be more limited and the gating less steep.

Laser gain material	ML technique	τ_p (ps)	λ_0 (nm)	P_{out} (mW)	P_{peak} (W)
Er^{3+} fluoride fiber	Fe:ZnSe	~19	2783	51.4	49
	SESAM	25	2780	1050	1900
	Graphene	42	2784.5	18	17
	SESAM	60	2797	440	140
	NPR	0.207	2805	44	3500
	NPR	0.497	2793	206	6400
	NPR	0.131	2780	317	23,000
	BP	42	2783	613	610
	BP	—	2771	6.2	...
	NLM	—	~2800	88	...
Ho^{3+}/Pr^{3+} fluoride fiber	SESAM	24	2870	132	200
	AOM	5	2870	34	330
	InAs SA	6	~2860	69.2	470
	SESAM	6.4	2710–2820	203	1100
	Bi_2Te_3	~6	2830	90	1400
	BP	8.6	2866.7	87.8	730
	NPR	0.180	2876	327	37,000
	Cd_3As_2 film	6.3	2860	...	...
Dy^{3+} fluoride fiber	FSF	33	2970–3300	120	82
	NPR	0.724	3050	175	3110
Er^{3+} fluoride fiber	BP	...	3462	40	...

Table 3.
Mode-locked Er^{3+}-, Ho^{3+}/Pr^{3+}, and Dy^{3+}-doped fluoride fiber lasers in the 2.8 μm–3.5 μm spectral region. [Reprinted/Adapted] with permission from [14].

4.3.2 Passive mode-locking

Passive mode-locking uses saturable absorbers such as SESAMs or nonlinear polarization evolution. In SESAM-based designs, the cavity might contain a short free-space section or a fiber-integrated arrangement that directs the intracavity beam onto a semiconductor mirror whose reflection is intensity-dependent. Low-intensity light is partially absorbed, while high-intensity pulses experience less absorption, thus reinforcing the growth of short, intense pulses. Nonlinear polarization evolution is a purely fiber-based scheme, relying on the intensity-induced changes in polarization state, followed by a polarizer that imposes intensity discrimination. These passive schemes can generate pulses without needing synchronous electrical signals.

Figure 9 shows two different techniques for mode-locking in lasers: active mode-locking (a) and passive mode-locking (b), both used to generate ultrashort pulses by enforcing phase coherence among longitudinal modes of the laser cavity. In active mode-locking (a), an external modulator—such as an acousto-optic or electro-optic modulator—is placed inside the laser cavity and driven at a modulation frequency f_m that matches the cavity mode spacing Δf_c. This periodically modulates the intracavity loss or phase, enforcing a regular train of short pulses with a repetition rate

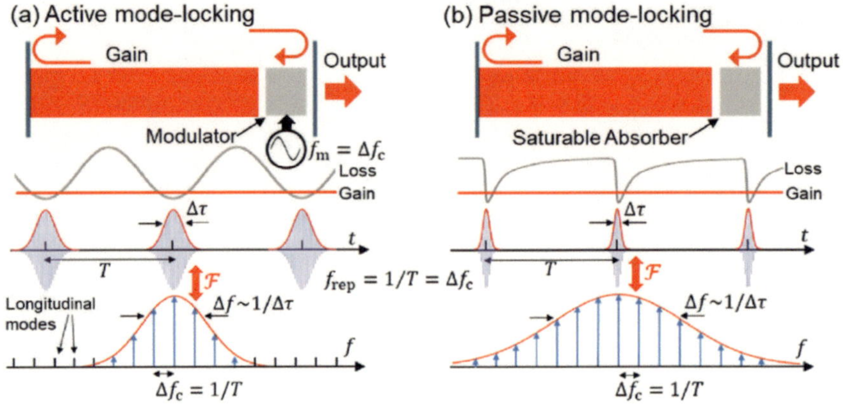

Figure 9.
Mode-locked laser. (a) Active modeo.131-locking and (b) passive mode-locking. [Reprinted/Adapted] with permission from [15].

$f_{rep} = 1/T = \Delta f_c$. The time-domain plot shows that pulses are evenly spaced by T, while in the frequency domain, the spectrum consists of discrete longitudinal modes spaced by Δf_c, with the pulse duration $\Delta\tau$ inversely proportional to the spectral bandwidth Δf. In passive mode-locking (b), no external modulation is needed; instead, a saturable absorber is introduced into the cavity. This element preferentially absorbs low-intensity light while allowing high-intensity pulses to build up, effectively shaping ultrashort pulses automatically. The resulting pulse train has the same periodicity T as in active mode-locking, but the spectral bandwidth Δf is typically broader, leading to much shorter pulses due to the inverse relationship $\Delta f \sim 1/\Delta\tau$. Passive mode-locking is the dominant method for generating femtosecond and picosecond pulses in ultrafast lasers.

Both methods enable stable short-pulse generation, but passive mode-locking is preferred for achieving the shortest pulse durations with high peak power, making it widely used in applications like ultrafast spectroscopy, precision machining, biomedical imaging, and telecommunications.

4.4 Nonlinear pulse formation

In fiber lasers, nonlinear effects play a critical role in shaping and stabilizing the pulses. Two of the most relevant phenomena are SPM and dispersion management. SPM arises from the Kerr effect, where the local refractive index changes proportionally to the pulse intensity. This change imparts a time-dependent phase shift on the pulse, often broadening its spectrum. If the dispersion in the fiber is suitably managed, the pulse can evolve into a soliton-like shape that remains stable over long propagation distances. In normal dispersion regimes, pulses can undergo parabolic or self-similar evolution, which can later be compressed outside the cavity to achieve very short pulse durations. Nonlinear pulse formation can also be harnessed to achieve supercontinuum generation, in which a pulse can be dramatically broadened in spectrum, providing broad bandwidth sources useful for spectroscopy and wavelength conversion.

4.4.1 Dispersion-managed soliton

Dispersion-managed soliton fiber lasers apply balanced segments of normal and anomalous dispersion fiber to support soliton or similar pulse shaping. In these systems, the nonlinear phase shift imparted to the pulse by SPM can be offset by the dispersive broadening or compression that occurs in each segment. The net result is a stable pulse that remains short over repeated cavity round trips.

Figure 10 shows the pulse evolution in a fiber laser cavity under different mode-locking regimes: soliton mode-locking (a), stretched pulse (SP) mode-locking (b), and dissipative soliton (DS) mode-locking (c). These regimes are distinguished by their dispersion management and pulse shaping mechanisms. In (a) soliton mode-locking, the pulse propagates in a fiber with anomalous dispersion, maintaining a constant duration τ_p. This results in a stable soliton due to the balance between dispersion and nonlinearity, leading to well-confined ultrashort pulses. The presence of an optical coupler (OC) and SA facilitates pulse shaping. In (b) stretched pulse mode-locking, the fiber cavity incorporates alternating sections of anomalous and normal dispersion, creating a periodic stretching and compression of the pulse. This dispersion-managed configuration enables higher pulse energies by mitigating nonlinear effects. The OC and SA again play a role in stabilizing the pulse. In (c) dissipative soliton mode-locking, the pulse propagates entirely in the normal dispersion regime, experiencing significant spectral broadening and pulse stretching. Unlike solitons, dissipative solitons are stabilized by energy dissipation mechanisms, including the spectral filter (SF), which enforces a stable pulse shape. This configuration allows for the generation of high-energy pulses due to strong chirping effects. This figure shows how dispersion management and additional optical components

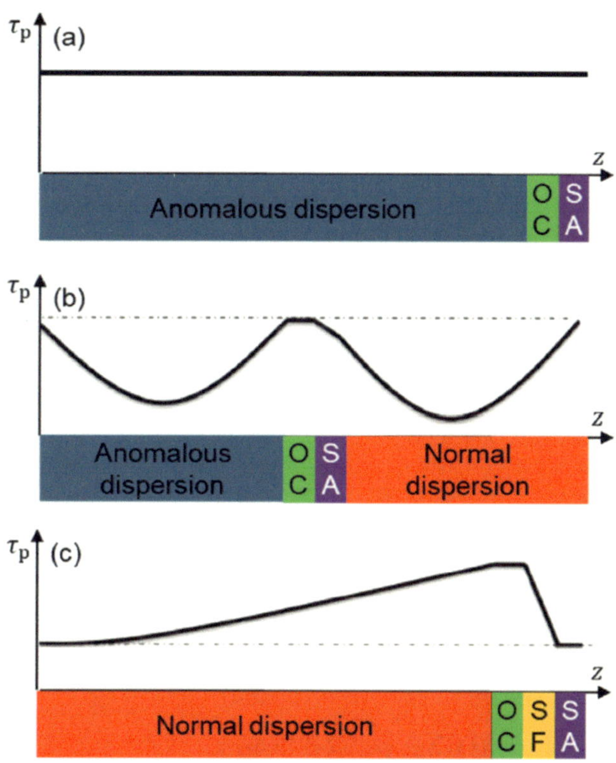

Figure 10.
Pulse evolution in fiber laser cavity. (a) Soliton mode-locking, (b) stretched pulse (SP) mode locking, and (c) dissipative soliton (DS) mode locking. OC: Optical coupler, SF: Spectral filter. [Reprinted/Adapted] with permission from [15].

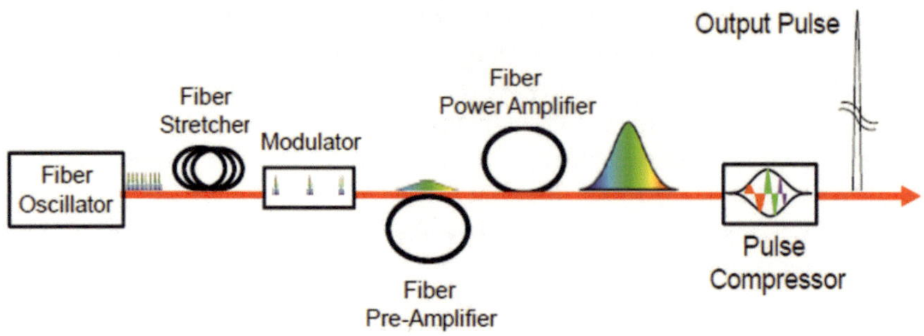

Figure 11.
Schematic of fiber-chirped pulse amplification. [Reprinted/Adapted] with permission from [16].

influence pulse evolution in fiber lasers, enabling different mode-locking techniques for various ultrafast laser applications.

4.4.2 Chirped pulse amplification

Chirped pulse amplification (CPA) is essential for applications requiring high-energy femtosecond pulses. The initial low-energy seed pulses from a mode-locked oscillator are stretched in time to reduce their peak power, then amplified in a series of fiber amplifier stages, and finally recompressed to near their original duration by a dispersive compressor. This approach prevents excessive nonlinear effects or damage in the amplifier, while ultimately delivering extremely high peak powers at the output.

Figure 11 shows the fiber-chirped pulse amplification (FCPA) process, a widely used technique for generating high-energy ultrafast laser pulses. The system begins with a fiber oscillator, which generates an initial ultrashort pulse. However, to prevent nonlinear effects and damage during amplification, the pulse is first stretched in time using a fiber stretcher, which introduces chirping—spreading the pulse's spectral components in time. A modulator is then used to shape or control the pulse parameters before amplification. The chirped pulse is then passed through a fiber pre-amplifier, where it undergoes an initial gain stage to increase its energy. This is followed by a fiber power amplifier, where the pulse energy is further boosted while maintaining its stretched duration. As the pulse gains energy, its spectral bandwidth remains broad, allowing for high peak power after compression. Finally, the amplified chirped pulse is sent through a pulse compressor, which compensates for the initial chirping, realigning the spectral components to generate an ultrashort, high-intensity output pulse. This method is essential in applications requiring high-peak-power femtosecond or picosecond pulses, such as precision machining, biomedical imaging, and ultrafast spectroscopy. The fiber-based implementation offers advantages such as compactness, high efficiency, and robust beam quality, making FCPA a key technology in modern laser systems.

5. Applications of optical fiber pulsed lasers

Optical fiber pulsed lasers have rapidly permeated many industries and research fields. Their combination of compactness, high reliability, and beam quality makes them appealing for commercial products and large-scale deployments.

5.1 Material processing

Material processing significantly benefits from the high peak power and superior beam-focusing capabilities of fiber-based pulsed lasers. Nanosecond Q-switched fiber lasers are extensively employed for precision drilling of small holes in metals and polymers, as well as for marking and engraving applications. Typical nanosecond pulsed lasers used for metal engraving range from 10 to 50 nanoseconds with peak powers ranging from tens of kilowatts to several megawatts. In precision micromachining, femtosecond mode-locked fiber lasers, with pulse durations typically under 500 femtoseconds and peak powers reaching hundreds of megawatts, ensure minimal heat-affected zones, precise ablation thresholds, and fine control over feature dimensions [17]. These lasers are particularly crucial in the microelectronics industry, medical device manufacturing, and the production of high-value engineering components, offering unmatched precision compared to traditional mechanical methods.

5.2 Medical applications

Fiber-based pulsed lasers have become indispensable in medical applications, ranging from minimally invasive laser surgeries to advanced biomedical imaging. The ultrafast nature of femtosecond pulses significantly reduces thermal damage during tissue ablation, enabling safer and more precise surgical procedures. In ophthalmology, femtosecond fiber lasers operating around 1030 nm have revolutionized corneal surgeries, significantly reducing procedure time, healing duration, and improving overall patient outcomes compared to conventional excimer lasers [18]. Furthermore, multiphoton imaging techniques utilizing ultrafast fiber lasers have transformed bioimaging, allowing deeper penetration into living tissue (up to several hundred micrometers) with significantly reduced phototoxicity and photobleaching. Additionally, fiber lasers doped with rare-earth elements such as Thulium and Holmium emit at eye-safe wavelengths around 2 μm, providing increased safety for patients and practitioners alike.

5.3 Telecommunications and optical communication

In telecommunications, erbium-doped fiber amplifiers (EDFAs), operating predominantly around 1.55 μm, form the backbone of long-haul optical communication networks. Pulsed fiber lasers have advanced significantly as coherent optical sources for high-speed modulation formats, such as quadrature amplitude modulation (QAM) and phase-shift keying (PSK), critical for ultra-high-capacity, long-distance data transmission. Time-division multiplexing (TDM) networks have extensively adopted pulsed sources for synchronization and high-speed switching applications. Furthermore, fiber-based lasers serve effectively as pump sources in Raman amplification schemes, enhancing signal-to-noise ratios and enabling seamless WDM [19].

5.4 LiDAR and remote sensing

Light Detection and Ranging (LiDAR) and remote sensing technologies extensively employ pulsed fiber lasers due to their precise temporal control and robust physical characteristics. Pulses typically in the range of nanoseconds to picoseconds enable accurate distance measurements with resolutions often better than 1 cm. Specific wavelength selections, such as 1.55 μm, are advantageous for eye-safe LiDAR used in autonomous vehicles and aviation, while Doppler LiDAR systems leverage wavelength stability and spectral purity for wind velocity and aerosol detection [20]. Fiber lasers' inherent resilience to environmental stress and ease of integration into compact packages enable their deployment in harsh operational environments, such as wind farms, aerial surveying platforms, and automotive applications, where reliability and accuracy are paramount.

5.5 Scientific applications

Fiber lasers play a critical role in various advanced scientific research fields, including frequency metrology, precision spectroscopy, and ultrafast dynamics studies. Mode-locked fiber lasers producing frequency combs have revolutionized optical frequency metrology, significantly enhancing atomic clocks' precision, with frequency uncertainty now approaching 10^{-19} levels [21]. Precision spectroscopy techniques utilize fiber lasers' wide tunability and spectral purity to probe atomic and molecular transitions with unprecedented resolution. Furthermore, ultrafast pulse durations in the femtosecond regime allow real-time studies of fundamental processes in chemistry, biology, and physics, elucidating reaction dynamics and molecular interactions on ultrashort timescales. Over the past decade, advancements in doped fiber technologies have progressively increased achievable average power levels (now routinely reaching several tens to hundreds of watts), pulse energies (microjoules to millijoules), and extended spectral coverage from visible to mid-infrared wavelengths, cementing fiber lasers as essential tools in contemporary scientific inquiry.

6. Conclusions

The field of optical fiber pulsed lasers has blossomed into one of the most dynamic and successful branches of photonics. By engineering the composition and structure of doped fibers, selecting appropriate pulse generation methods such as gain-switching, Q-switching, or mode-locking, and leveraging sophisticated dispersion and nonlinear management, laser designers can achieve pulsed outputs from nanoseconds down to femtoseconds with exceptional beam quality. Fiber architectures bring inherent alignment stability, compactness, and scalability, making these systems suitable for both cutting-edge scientific experiments and demanding industrial tasks. The material processing sector benefits from versatile fiber lasers that can drill, engrave, or cut with minimal thermal damage. Medical procedures ranging from laser surgery to high-resolution imaging exploit the precise temporal and spatial characteristics of fiber-based pulsed sources. Next-generation communications and advanced sensing, including LiDAR and environmental monitoring, rely on fiber laser technology for robust, high-speed, and spectrally flexible performance.

Challenges continue to emerge when scaling to higher powers and energies, or when pushing for ultra-broad bandwidths and extremely short pulses. Nonetheless, active research efforts in fiber design, doping chemistry, thermal management, and nonlinear control have repeatedly demonstrated that fiber lasers can adapt to meet these demands. It is reasonable to expect that integrated photonic solutions, quantum-compatible devices, and mid-infrared or ultraviolet fiber lasers will open new vistas in the coming years. By merging the latest advances in materials science, optics, and engineering, fiber pulsed laser systems are positioned to remain a vital, evolving toolset for science, industry, and society at large.

Acknowledgements

The authors declare that there are no personal acknowledgments or funding sources to report for this work.

Conflict of interest

The authors declare no conflict of interest.

Author details

Wei Zhang
Baylor University, Waco, TX, USA

*Address all correspondence to: wei_zhang3@baylor.edu

References

[1] Zervas MN, Codemard CA. High power fiber lasers: A review. IEEE Journal of Selected Topics in Quantum Electronics. 2014;**20**(5):219-241. DOI: 10.1109/JSTQE.2014.2321279

[2] Nilsson J, Payne DN. High-power fiber lasers. Science. 2011;**332**:921-922. DOI: 10.1126/science.1194863

[3] Heebner JE, Wong V, Schweinsberg A, Boyd RW, Jackson DJ. Optical transmission characteristics of fiber ring resonators. IEEE Journal of Quantum Electronics. 2004;**40**(6):726-730. DOI: 10.1109/JQE.2004.828232

[4] Ying D, Ma H, Jin Z. Ringing phenomenon of the fiber ring resonator. Applied Optics. 2007;**46**(22):4890-4895. DOI: 10.1364/AO.46.004890

[5] Zhang W, Menyuk CR, Hu J. A study of coupling efficiency in pulley microring resonators. In: Proceedings of the Frontiers in Optics + Laser Science (FIO, LS); 17-20 October 2022; Rochester, New York, United States. Washington, DC: Optica Publishing Group; 2022. Paper JTu5A.52

[6] Hercher M. An analysis of saturable absorbers. Applied Optics. 1967;**6**:947-954. DOI: 10.1364/AO.6.000947

[7] Keller U, Weingarten KJ, Kopf D, Braun B, Jung I-D, Fluck R, et al. Semiconductor saturable absorber mirrors (SESAMs) for femtosecond to nanosecond pulse generation in solid-state lasers. IEEE Journal of Selected Topics in Quantum Electronics. 1996;**2**(3):435-453. DOI: 10.1109/2944.571743

[8] Martinez A, Sun Z. Nanotube and graphene saturable absorbers for fibre lasers. Nature Photonics. 2013;**7**:842-845. DOI: 10.1038/nphoton.2013.304

[9] Zhang W, Gattass RR, Shaw LB, Menyuk CR, Hu J. Maximizing the FWM conversion efficiency in a PCF with a randomly varying pitch. Journal of the Optical Society of America B: Optical Physics. 2024;**41**:2408-2413. DOI: 10.1364/JOSAB.537104

[10] Swiderski J. Recent development of mid-infrared supercontinuum generation in fluoroindate glass fibers. Applied Sciences. 2022;**12**(10):4927. DOI: 10.3390/app12104927

[11] Yang J, Tang Y, Xu J. Development and applications of gain-switched fiber lasers [invited]. Photonics Research. 2013;**1**:52-57. DOI: 10.1364/PRJ.1.000052

[12] Chen X, Wang N, He C, Lin X. Development of all-fiber nanosecond oscillator using actively Q-switched technologies and modulators. Optics & Laser Technology. 2023;**157**:108709. DOI: 10.1016/j.optlastec.2022.108709

[13] Qi Y, Bai Z, Wang Y, Zhang X, Qi Y, Ding J, et al. Research progress of all-solid-state passively Q-switched Er:Yb:Glass lasers. Infrared Physics & Technology. 2021;**116**:103727. DOI: 10.1016/j.infrared.2021.103727

[14] Ma J, Qin Z, Xie G, Qian L, Tang D. Review of mid-infrared mode-locked laser sources in the 2.0 μm–3.5 μm spectral region. Applied Physics Reviews. 2019;**6**(2):021317. DOI: 10.1063/1.5037274

[15] Yamashita S, Set SY. Advances and challenges of mode-locked fiber lasers. Optics Communications.

2025;**578**:131406. DOI: 10.1016/j.optcom.2024.131406

[16] Cho GC, Liu B, Shah L, Liu Z, Che Y, Xu J. Latest developments of ultrafast fiber laser and its material applications. In: Proceedings of SPIE 7214, Ultrafast Phenomena in Semiconductors and Nanostructure Materials XIII; 6 February 2009; San Jose, California, United States. Bellingham, WA: SPIE; 2009. Paper 72140R

[17] Dondieu SD, Wlodarczyk KL, Harrison P, Rosowski A, Gabzdyl J, Reuben RL, et al. Process optimization for 100 W nanosecond pulsed fiber laser engraving of 316L grade stainless steel. Journal of Manufacturing and Materials Processing. 2020;**4**(4):110. DOI: 10.3390/jmmp4040110

[18] Crotti C, Deloison F, Alahyane F, Aptel F, Kowalczuk L, Legeais JM, et al. Wavelength optimization in femtosecond laser corneal surgery. Investigative Ophthalmology & Visual Science. 2013;**54**:3340-3349. DOI: 10.1167/iovs.12-10694

[19] Nakazawa M. Evolution of EDFA from single-core to multi-core and related recent progress in optical communication. Optical Review. 2014;**21**:862-874. DOI: 10.1007/s10043-014-0139-1

[20] Li N, Ho CP, Xue J, Lim LW, Chen G, Fu YH, et al. A progress review on solid-state LiDAR and nanophotonics-based LiDAR sensors. Laser & Photonics Reviews. 2022;**16**:2100511. DOI: 10.1002/lpor.202100511

[21] Ma LS, Bi Z, Bartels A, Robertsson L, Zucco M, Windeler RS, et al. Optical frequency synthesis and comparison with uncertainty at the 10^{-19} level. Science. 2004;**303**(5665):1843-1845. DOI: 10.1126/science.1095092

Chapter 2

Rotating Mirror Q-Switched CO Laser: Optimization of Power Spectrum on Mirror Rotation – A Theoretical Study

Utpal Nundy, Ramesh Chandra Das and Ayan Ghosh

Abstract

This theoretical study explores the performance of a cryogenically cooled Q-switched CO laser utilizing a Rotating Mirror Q-Switching technique. By varying the mirror rotation frequency from 50 to 625 Hz, we optimized the output power and spectral characteristics, achieving a peak power of 200 kW at 500 Hz with a pulse FWHM of 220 ns. The Q-switched pulse comprises approximately 100 CO laser lines spanning 5.0–6.2 μm, with 65 lines exceeding 100 W and a maximum of 25 kW. These characteristics make the laser highly suitable for nonlinear sum-frequency generation in crystals. This work provides novel insights into the operational physics of Q-switched CO lasers and their potential for advanced spectroscopic applications.

Keywords: Rotating Mirror Q-Switching, CO laser, power spectrum of CO laser, theoretical modeling of CO laser, mid-infrared (mid-IR) laser source, laser performance optimization, nonlinear frequency generation

1. Introduction

The mid-infrared (mid-IR) region hosts two significant laser systems: the carbon dioxide (CO_2) laser, operating in the 9–11 μm region, and the carbon monoxide (CO) laser, operating in the 5–7 μm region. Both systems deliver high power and good efficiency, making them valuable for diverse applications. CO_2 lasers are widely used in industrial settings for material cutting, welding, and surface processing due to their reliability and ease of operation. In contrast, CO lasers, despite their potential, are less common, primarily due to the complexity of cryogenic cooling required for their operation. However, CO laser lines exhibit superior absorption by structural materials and biological tissues, positioning them as promising alternatives to CO_2 lasers for advanced material processing and biomedical applications. Additionally, CO lasers are highly effective for pollution monitoring and atmospheric sensing, as their emission lines are readily absorbed by many organic and inorganic compounds [1].

The distinct spectral characteristics of these lasers stem from their molecular structures. The CO_2 molecule has three vibrational modes, with laser transitions occurring between vibrational levels of two modes, typically resulting in single-line emission at 10.6 μm (P-20) or adjacent lines (e.g., P-18, P-22) in both continuous-wave (CW) and pulsed modes. Conversely, the CO molecule has a single vibrational mode, forming a simple energy ladder that naturally produces multiline emission unless constrained to a single line using a grating. Multiline operation in CO_2 lasers, achieved through specialized techniques, is critical for applications such as laser fusion and isotope enrichment [2].

Among various techniques of Q-switching [3], Rotating Mirror Q-Switching (RMQS) stands out for its simplicity and popularity. Osgood et al. [4] were among the pioneering researchers to build a CW CO laser and successfully Q-switch it. Osgood reported the observation of 20 spectral lines during CW operation and 62 lines during Q-switched operation. In the Q-switched mode, the laser spectrum extended from 5 to 6.2 μm. Furthermore, the wavelength range and the number of lines can be extended through nonlinear frequency generation, such as sum-frequency generation using nonlinear crystals like $ZnGeP_2$ [5]. Such broad-spectrum sources find important applications in spectroscopy and chemical analysis [6, 7].

In the nonlinear interaction process, pulses interact in pairs; thus, both the number of lines and their intensities play critical roles in efficient frequency generation. Consequently, a higher peak power not only improves intensity but also supports a greater number of active spectral lines. Hence, optimizing the laser spectrum is closely tied to optimizing the peak power. However, as we will demonstrate, the composite Q-switched output pulse is composed of three distinct types of sub-pulses: Some start early but rapidly decay. Others initiate early and persist through the pulse duration—these are responsible for achieving the highest intensities and dominate the laser output. A third category emerges later, contributing mainly to the pulse tail.

Time synchronization among these pulses is critical for efficient sum-frequency generation. This topic was discussed in detail in an earlier publication [8] and will only be discussed in brief here.

The laser spectra and power output of a room temperature CO laser with mirror rotation were earlier studied experimentally by Gutin [9]. To the best of our knowledge, our work represents the first attempt to optimize a cryogenically cooled, Rotating Mirror Q-Switched CO laser.

In this study, we present a theoretical investigation of the variation in peak power and laser spectra for a system similar to that described by Osgood [4]—hereafter referred to as the OSGOOD laser. Section 2 describes the laser system and experimental setup. Section 3 discusses key physical processes relevant to RMQS. Section 4 outlines the methodology employed in our theoretical modeling; further details can be found in Ref. [8]. Section 5 presents the simulation results. Section 6 provides the conclusions, with emphasis on the implications for nonlinear frequency generation applications.

2. Experimental setup

A schematic diagram of the experimental setup is shown in **Figure 1**. For our modeling, we have chosen a laser similar to the OSGOOD laser. The parameters of this laser are presented in a compact form in **Table 1**. The laser employs a longitudinal

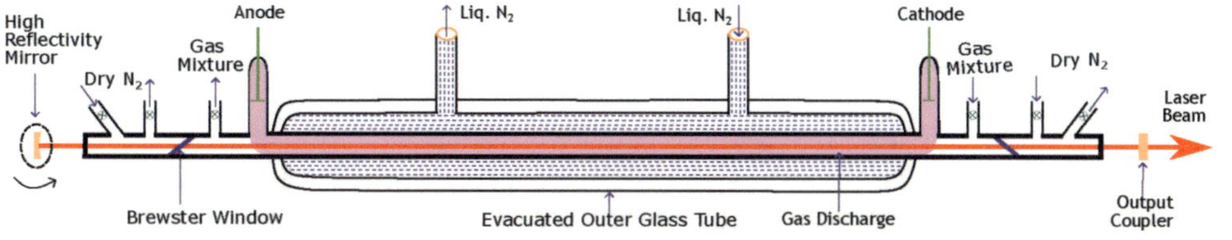

Figure 1.
Schematic of experimental setup.

Gas mixture and pressure (Torr)	Discharge length (cm)	Discharge Dia. and mirror Dia. (mm)[a]	Gas temperature (K)	OC reflectivity (%)	Cavity length (cm)	Laser beam Dia. (mm)	Discharge voltage (kV) and current (mA)
CO: 0.2, He: 6.0	120	40	167.5	0.9	200	26.6	9 kV, 20 mA

[a]*Both the diameters are same.*

Table 1.
Parameters of the laser system.

discharge tube cooled by liquid N_2, with electrical excitation provided by a DC discharge. To ensure thermal insulation, the outer chamber surrounding the discharge tube is evacuated. Brewster windows are used at both ends of the tube to minimize the reflection losses. The laser resonator consists of a fully reflecting curved mirror with a radius of curvature of 10 meters and a partially reflecting plane output coupler. Q-switching is achieved by rotating the curved mirror at various frequencies, ranging from 50 to 625 Hz in our case. For this cavity configuration, the laser mode was calculated [10] to be a multimode beam with a diameter of 26.6 mm.

3. Physical processes

3.1 Changing laser mode volume

When the optical cavity is made up of a curved mirror and a plane mirror, the laser mode is centered along the optic axis. The optic axis goes through the centers of both the mirrors and is perpendicular to them. As the curved mirror rotates, the optic axis moves parallel to itself over the surfaces of both the mirrors to remain perpendicular to the mirror surfaces. The laser mode also traverses along the mirror surface. This is shown in **Figure 2**.

The laser mode, initially centered at point "O," enters the left edge of the mirror and moves toward the right. At this stage, only approximately half of the mode area lies within the mirror surface. As it continues moving rightward and reaches point "A," the laser mode achieves its full volume. During this transition from the left edge to point "A," the mode volume progressively increases. The full mode volume is maintained as the center "O" travels from point "A" to point "B." Beyond point "B," as "O" moves further to the right, the mode volume decreases, reducing to half of its full value when it reaches point "C." Let the times taken to reach points "A," "B," and "C"

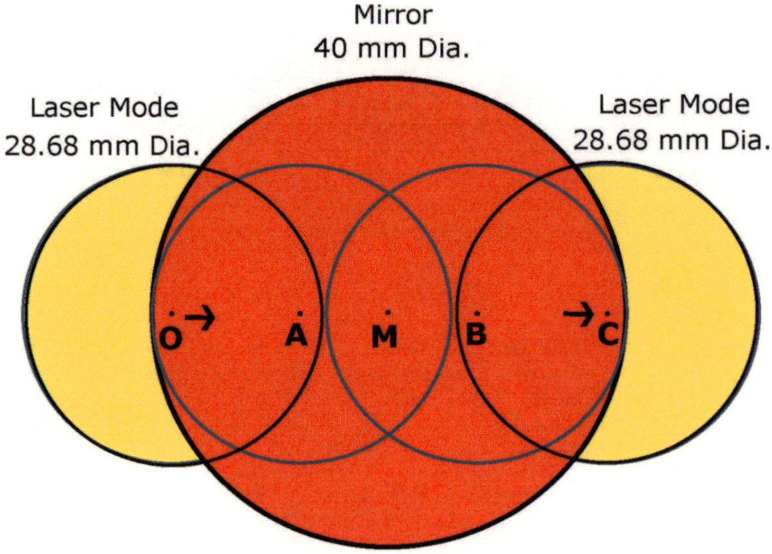

Figure 2.
Movement of the laser mode on the mirror surface.

be denoted by T_A, T_B, and T_C respectively. **Table 2** summarizes the values of T_A, T_B, and T_C for various mirror rotation frequencies.

3.2 Laser cavity loss variation with mirror rotation

The laser mode has a Gaussian distribution of its intensity over space. The laser radius is an approximation and is taken as the distance from center where the intensity has dropped to e^{-2} of the peak intensity. When the mode moves along the mirror surface, a part of this distribution is intercepted by the mirrors. It is obvious that the interception factor is maximum when the center of the laser mode coincides with the center of the mirror. This interception factor can be calculated by the "Cumulative Distribution Function (CDF)," which is tabulated in Ref. [11]. **Figure 3** shows the combined interception factors of the two mirrors with respect to time for mirror rotation frequency of 200 Hz.

Frequency (Hz)	Mode travel velocity (m/s)[a]	T_A (ns)	T_B (ns)	T_C (ns)
50	3125	4267	8533	12,800
100	6250	2133	4266	6400
125	7812.5	1706	2413	5120
200	12,500	1066	2123	3200
400	25,000	533	1066	1600
500	31,250	427	853	1280
625	39,062.5	341	682	1024

[a]*This refers to the linear speed (in m/s) at which the laser mode traverses the mirror surface.*

Table 2.
Laser mode velocity and values of TA, TB, and TC with mirror rotation frequency.

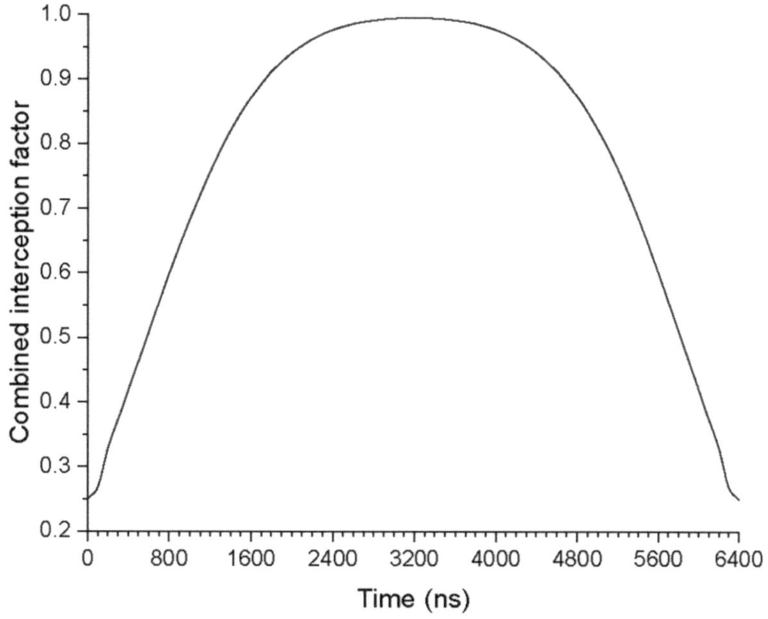

Figure 3.
Combined interception factor for mirror rotation frequency of 200 Hz. The curve is similar for other frequencies. The time scale is proportionately shorter at higher frequencies and longer for lower frequencies.

4. Calculation methodology

We have chosen the OSGOOD laser for our modeling work, as it has been theoretically studied by several authors [12, 13], and the fundamental data required for modeling is readily available. The CO molecule serves as an example of an anharmonic oscillator, characterized by a vibrational energy level ladder. Laser transitions occur among the rotational sublevels associated with these vibrational levels. The CO laser operates exclusively on P-branch transitions [1].

In our model, we consider 18 vibrational bands, ranging from levels 22 to 5, and 8 rotational transitions, P7 to P14, within each band. We adopt the following notation: a laser line is designated as VPJ, where V denotes the vibrational band (i.e., the quantum number of the upper vibrational level), and rotational transitions occur between sublevels $J - 1$ and J, corresponding respectively to the upper and lower vibrational levels. For instance, 22P7 refers to a transition between the levels 22 ($J = 6$) and 21 ($J = 7$).

Our model is based on a rate equation approach. It tracks the population densities of various vibrational levels, the rotational sublevel population densities corresponding to 144 laser lines, and the photon densities associated with these lines. The rate equations describe the kinetic processes that govern changes in these population densities. For vibrational levels, the changes arise from electrical pumping, spontaneous emission, collisional deactivation due to "He" in the gas mixture, upward and downward flux due to CO-CO collisions (anharmonic pumping), and laser-induced interactions. For rotational sublevels, we account for rotational relaxation, assuming a rotational relaxation time of 50 ns.

For photon densities, growth due to population inversion and loss through cavity decay (including losses described in Section 3) are considered. Altogether, the system consists of 364 coupled nonlinear equations—one for each parameter discussed. The initial values of all parameters are calculated and known before the integration begins.

These nonlinear equations are integrated over time in steps of 0.5 ns using an embedded fifth-order Runge-Kutta algorithm as prescribed by Press et al. [14]. Referring to **Figure 2**, the calculations begin when the laser mode center "O" is at the left edge of the mirror and continue until "O" reaches the right edge of the mirror at point "C."

5. Results

We have varied the rotation frequency of the mirror from 50 to 625 Hz and calculated the intensities of the laser pulses. The calculations automatically provide the laser spectrum. **Table 3** below summarizes the main features of these Q-switched laser pulses. As the frequency increases, the peak power of the Q-switched pulse rises and the pulse becomes narrower. However, beyond 500 Hz, the peak power begins to decrease. Referring to **Table 2**, the time T_C represents the total interaction time and can be considered the Q-switching time. The reduction in peak power beyond 500 Hz occurs because the delay time becomes a significant fraction of the Q-switching time, causing the laser pulse to grow while losses are simultaneously increasing [9]. Pulse termination is governed by gain saturation due to the photon density, and in most cases, the pulse duration is considerably shorter than the Q-switching time, except at 500 and 625 Hz where they become comparable. **Figure 4** shows the Q-switched pulses for 100, 125, and 200 Hz, while **Figure 5** presents the pulses at 400, 500, and 625 Hz. For consistency and ease of comparison, in all cases the pulses are plotted starting from the point where they begin to grow, omitting the delay time.

All these pulses are composite in nature, resulting from the superposition of intensities from lasing lines across different vibrational bands. Each individual pulse starts at a different time, peaks at a different time, and decays differently. Consequently, the laser spectrum continuously evolves throughout the duration of the laser pulse.

Frequency (Hz)	Peak power (kW)	Time to peak (ns)	Delay (ns)[a]	Pulse FWHM (ns)	Pulse termination time (ns)
50	102.5	2900	2240	370	[b]
100	146.8	1850	1435	358	3200
125	157.8	1600	1250	330	2900
200	164.0	1200	955	320	2200
400	177.1	960	710	304.5	1400
500	208.5	880	660	223	1240
625	196.3	820	610	176	1000

[a]*From 0% to 10% of the peak intensity.*
[b]*The tail part is oscillatory, reaches a minimum of 0.9 kW at 3800 ns, grows to a second peak of 1.3 kW at 4100 ns, and then the intensity decays to 0 at 10,000 ns.*

Table 3.
Pulse characteristics at different frequencies.

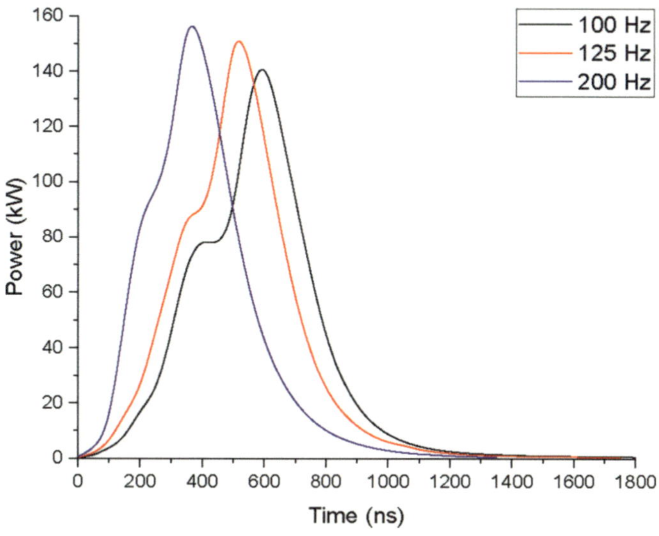

Figure 4.
Q-switched pulses at frequencies 100, 125, and 200 Hz.

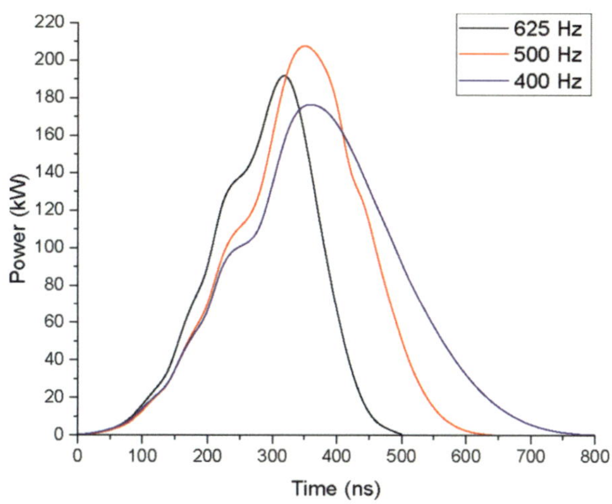

Figure 5.
Q-switched laser pulses at frequencies 400, 500, and 625 Hz.

The laser spectrum at the peak of the pulse—occurring at 880 ns for f = 500 Hz—is shown in **Figure 6**. The intensities of the individual lasing lines span a wide range, from a few milliwatts to several tens of watts.

The Q-switched laser pulse is shaped predominantly by the contributions of approximately 100 individual laser pulses, each operating with different delays, peak intensities, and times to reach their maximum. Among these, the lines 12P9, 12P10, 11P9, 11P10, 11P11, 10P9, 10P10, 9P9, and 9P10 consistently begin lasing first across all cases. We refer to these nine lines as the seed pulses. These seed pulses play a crucial role in determining the peak power and shape of the Q-switched pulse. Although their intensities vary, they are synchronized—starting to grow simultaneously, peaking close in time, and decaying together. These seed pulses initiate a cascade lasing

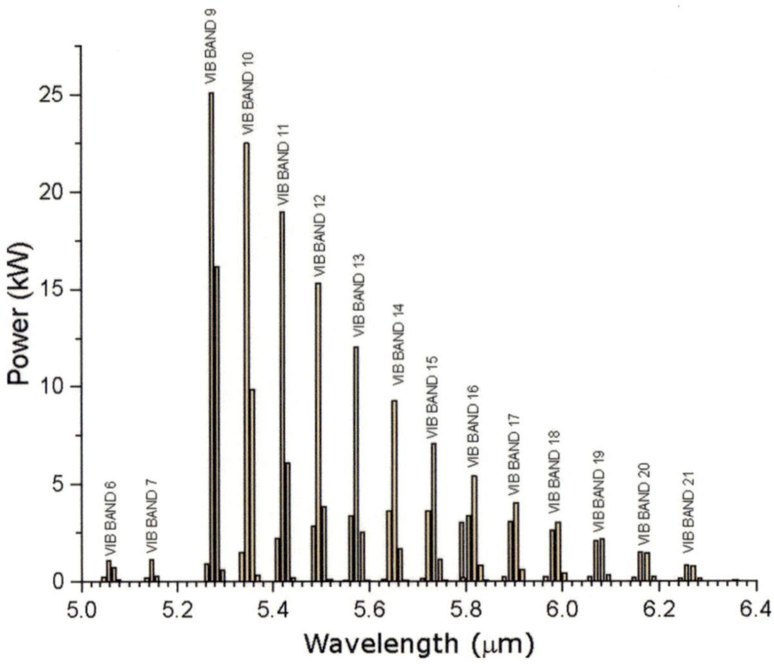

Figure 6.
The laser spectrum at the peak of the pulse at 880 ns. The intensities for Vib Bands 5, 8, and 22 are too low to show up in this figure.

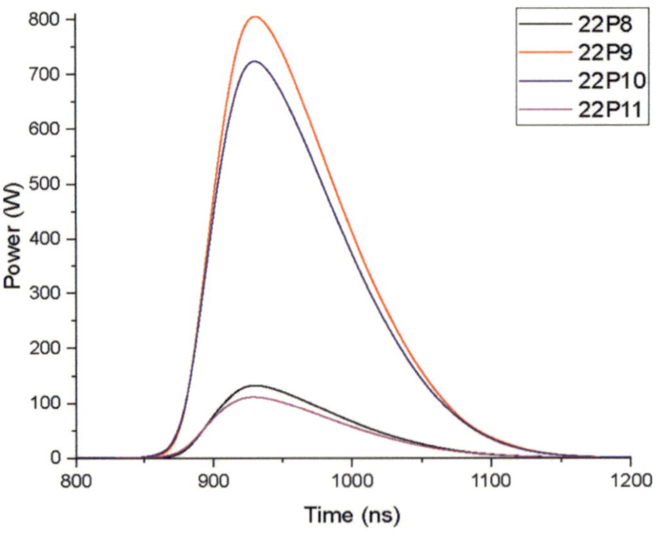

Figure 7.
Lasing in Vibration Band 22.

process [1], sequentially inducing lasing in both upper and lower vibrational bands. The evolution of laser pulses in various vibrational bands is illustrated as follows: Vibration Band 22 in **Figure 7**, Vibration Band 14 in **Figure 8**, Vibration Band 9 in **Figure 9**, Vibration Band 8 in **Figure 10**, and Vibration Band 5 in **Figure 11**.

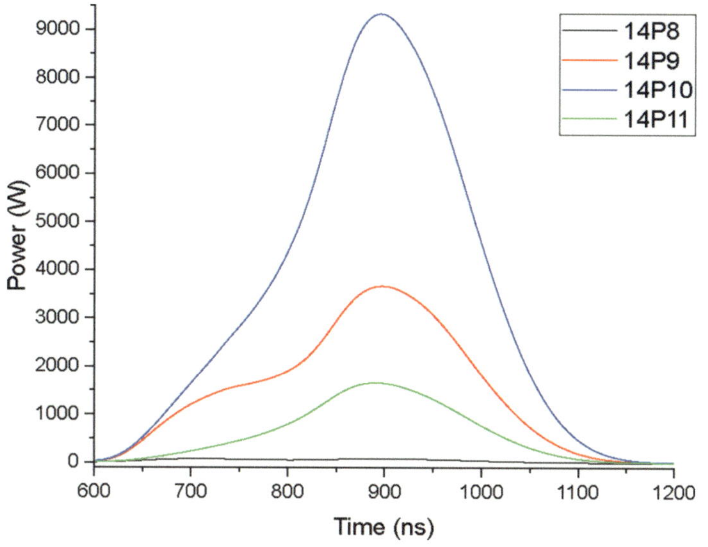

Figure 8.
Lasing in Vibration Band 14.

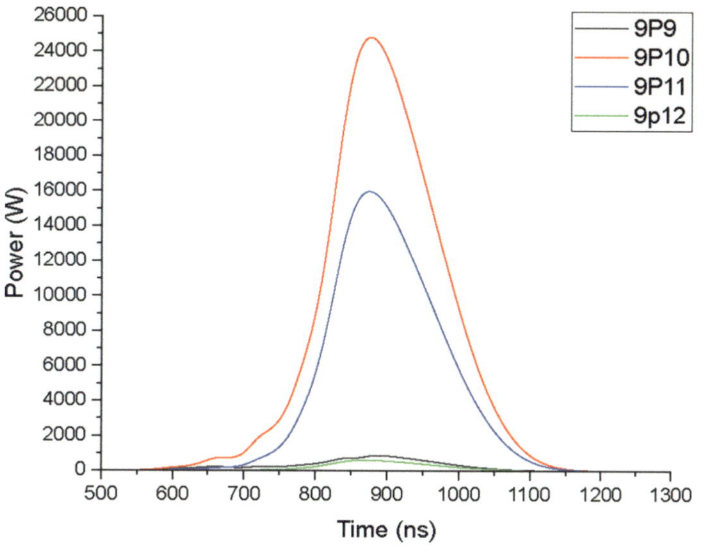

Figure 9.
Lasing in Vibration Band 9.

The Q-switched laser pulse peaks at 880 ns and extends up to 1240 ns. Among the various vibrational bands, lasing in Vibration Bands 22 and 5 exhibits a noticeable delay and primarily contributes to the trailing edge of the Q-switched pulse. In contrast, lasing in Vibration Band 14 occurs with only a small delay relative to Vibration Band 9, with both bands peaking around the same time and sustaining their lasing action throughout the duration of the Q-switched pulse. Lasing in Vibration Band 8, on the other hand, starts early but terminates before the peak of the Q-switched pulse, thus contributing mainly to its leading edge.

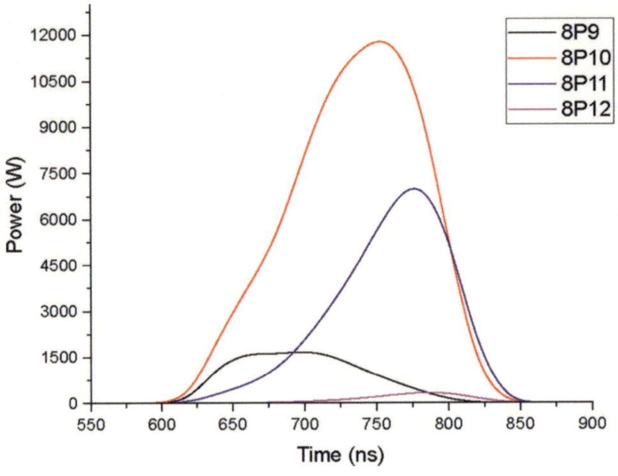

Figure 10.
Lasing in Vibration Band 8.

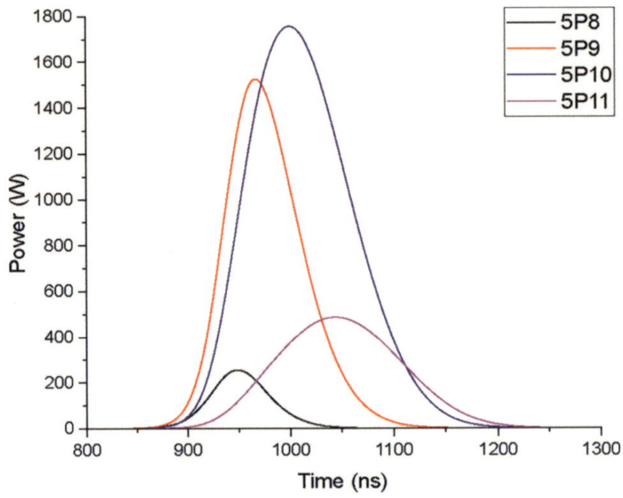

Figure 11.
Lasing in Vibration Band 5.

For effective sum-frequency generation (SFG) in a nonlinear crystal, two conditions are crucial: The pulses should have comparable intensities, and they must exhibit good temporal synchronization, or overlap, in time. The intensities of the lasing pulses vary widely, ranging from a few tens of milliwatts to several tens of kilowatts. By considering only those pulses with peak powers greater than 100 W, we find that 65 lasing lines fall within the range of 100 W to 25 kW.

To illustrate the issue of time synchronization, **Figure 12** shows the most intense pulses from Vibration Bands 22, 14, 9, 8, and 5, redrawn for clarity. It can be observed that the pulse from Vibration Band 9 overlaps well temporally with those from Bands 22, 14, and 5, suggesting strong potential for nonlinear interaction. However, the overlap with the pulse from Vibration Band 8 is comparatively poor, which would limit its contribution to efficient sum-frequency generation.

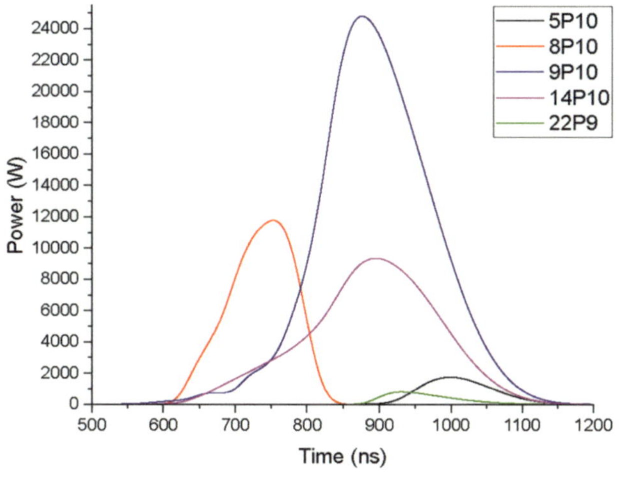

Figure 12.
Time synchronization between laser pulses in the Vibration Bands 22, 14, 9, 8, and 5.

In all cases discussed, the energy per Q-switched pulse is approximately 50 mJ. The average output power can be estimated by multiplying the pulse energy by the repetition frequency, f. Thus, at f = 100 Hz, the average power is approximately 5 W, and at f = 500 Hz, it increases to about 25 W.

6. Conclusions

In conclusion, this theoretical study optimizes a cryogenically cooled Q-switched CW CO laser, achieving a peak power of 200 kW at an optimal mirror rotation frequency of 500 Hz. The Q-switched pulse, comprising 65 high-power pulses, enables efficient sum-frequency generation in a $ZnGeP_2$ crystal, producing approximately 2000 laser pulses in the 2.5–3.0 μm wavelength range. The enhanced gain at liquid nitrogen temperature supports higher mirror rotation speeds, yielding pulses with greater peak power and narrower FWHM, thus improving frequency conversion efficiency. This work marks the first attempt to optimize a cryogenically cooled Rotating Mirror Q-Switched CO laser, providing valuable insights into its operational physics and paving the way for advanced nonlinear optical applications.

Author details

Utpal Nundy[1,2*], Ramesh Chandra Das[3,4] and Ayan Ghosh[3]

1 RRCAT, Indore, India

2 Kendriya Vihar, Kharghar, Navi Mumbai, India

3 Laser and Plasma Technology Division, Bhabha Atomic Research Centre, Mumbai, India

4 Homi Bhabha National Institute, Mumbai, India

*Address all correspondence to: unundy@yahoo.co.in

References

[1] Ionin AA. Electric discharge CO lasers. In: Endo M, Walter RF, editors. Gas Lasers. Boca Raton, London, New York: CRC Press; 2007. pp. 133-176

[2] Biswas DJ, Nath AK, Nundy U, Chatterjee UK. Multiline CO_2 lasers and their uses. In: Moss TS, editor. Progress in Quantum Electronics. Vol. 14. No. 1. Oxford, New York, Beijing, Frankfurt, Sao Paulo, Sydney, Tokyo, Toronto: Pergamon Press; 1991. pp. 1-42

[3] Biswas DJ. A beginner's guide to lasers and their applications—Part 1. In: Undergraduate Lecture Notes in Physics (Chapter 9). Switzerland: Springer Nature Switzerland AG; 2023. ISBN 2192-4791

[4] Osgood RM Jr, Nichols ER, Epperts WC Jr, Petty RD. Q-switching of the carbon monoxide laser. Applied Physics Letters. 1969;**15**(2):69-71

[5] Ionin AA, Kinyaevskiy IO, Klimachev YM, Mozhaeva VA, Andreev YM. Three-stage frequency conversion of sub-microsecond multiline CO laser pulse in a single $ZnGeP_2$ crystal. Optics Letters. 2018;**43**(13):3184-3187

[6] Ionin AA, Kinyaevskiy IO, Klimachev YM, Kozlov AY, Kotkov AA. A broadband laser source (2.5 to 17 μm) for plasma diagnostics. Physics of Atomic Nuclei. 2017;**80**(11):1635-1642

[7] Ionin AA, Kinyaevskiy IO, Klimachev YM, Sagitova AM, Andreev YM. CO laser sum frequency comb for atmosphere sensing. Infrared Physics and Technology. 2019;**100**:62-66

[8] Nundy U. Theoretical modelling of a rotating mirror Q-switched CW CO laser. Optics and Laser Technology. 2022;**146**:107466

[9] Gutin MA. Dependence of spectra and power of a sealed CO laser on the rate of Q-switching. Soviet Journal of Quantum Electronics. 1986;**16**(2):279-283

[10] Silfvast WT. Laser Fundamentals. South Asian Paperback Edition. Cambridge: Cambridge University Press; 2003

[11] Riley KF, Hobson MP, Bence SJ. Mathematical Methods for Physics and Engineering. Cambridge Low Price Edition. Cambridge: Cambridge University Press; 1998. p. 100

[12] Rich WJ. Kinetic modelling of the high-power carbon monoxide laser. Journal of Applied Physics. 1971;**42**(5):2179-2186

[13] Rockwood SD, Brau JE, Proctor WA, Canavan GH. Time-dependent calculations of carbon monoxide laser kinetics. IEEE Journal of Quantum Electronics. 1973;**QE-9**(1):120-126

[14] Press WH, Teukolsky SA, Vetterling WT, Flannery BP. Numerical Recipes in Fortran 77: The Art of Scientific Computing. 2nd ed. Cambridge: Cambridge University Press; 1992. p. 710

Chapter 3

Emerging Materials-Based Pulse Generation in Fiber Laser Systems

Xiangling Tian, Wenhao Wei, Peng Bi and Rongfei Wei

Abstract

Recent advances in low-dimensional nanomaterials have significantly expanded the landscape of saturable absorbers (SAs) for pulsed fiber laser systems, enabling efficient generation of ultrashort optical pulses through passive Q-switching and mode-locking. This chapter provides a comprehensive overview of the underlying physical mechanisms—such as saturable absorption, the Kerr effect, and nonlinear optical responses—and introduces widely used characterization techniques, including the Z-scan and double-balanced detection methods. Emphasis is placed on emerging functional materials, including zero-dimensional quantum dots, one-dimensional nanowires, and two-dimensional materials like graphene, transition metal dichalcogenides (TMDs), black phosphorus, topological insulators, and perovskites. The synthesis methods, optical properties, and fiber integration strategies of these SAs are discussed in detail. Representative case studies demonstrate their implementation in fiber laser systems across a broad spectral range—from visible to mid-infrared—with discussions on key performance metrics such as pulse width, repetition rate, output power, and operational stability. Finally, the chapter highlights current challenges and future directions in this rapidly evolving field, with applications spanning ultrafast photonics, optical communication, and sensing. This chapter aims to equip both researchers and engineers with a solid foundation for designing compact, cost-effective, and high-performance pulsed laser systems enabled by advanced nanomaterials.

Keywords: saturable absorber, pulse fiber laser, ultrafast photonics, low-dimensional materials, nonlinear optics, passive Q-switching and mode-locking, advanced functional materials

1. Introduction

The development of fiber lasers dates back to the 1960s, when trivalent rare-earth ions such as neodymium, erbium, and thulium were successfully incorporated into glass materials [1]. Pulsed laser systems are widely used across both fundamental research and industrial applications. In fiber lasers, two common techniques are employed to generate ultrashort pulses: Q-switching and mode-locking [2–4]. A typical ring cavity setup, illustrated in **Figure 1**, is often used due to its simplicity and efficiency. Each technique offers unique advantages that make it suitable for different applications. For instance, Q-switched lasers, known for their high peak power, are frequently utilized in laser machining and military technologies [4]. In contrast,

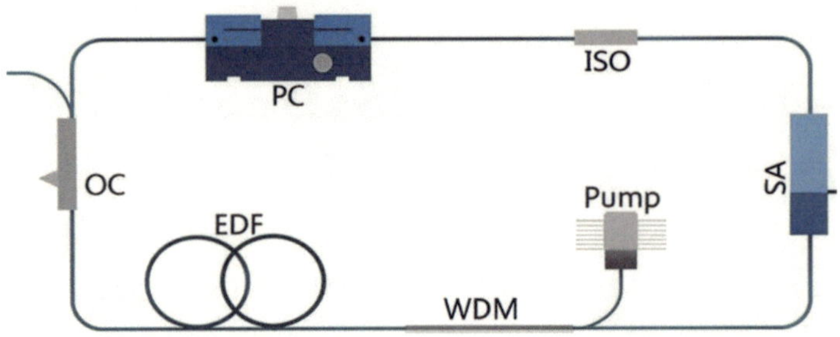

Figure 1.
Schematics of ring cavity setup for a mode-locked laser. In brief, WDM: wavelength division multiplexer, ISO: polarization independent isolator, PC: polarization controller, OC: optical coupler, EDF: erbium-doped fiber.

mode-locked lasers are ideal for precision applications, such as micromachining and nano-imaging, due to their ability to produce ultrafast pulses [4].

The evolution of saturable absorbers (SAs) has progressed from dye-based materials to more advanced technologies such as semiconductor saturable absorber mirrors (SESAMs) [2]. Following the emergence of graphene, there has been growing interest in developing SAs using novel functional materials. These include two-dimensional (2D) materials, such as graphene, topological insulators (TIs), layered transition metal dichalcogenides (TMDs), and black phosphorus (BPs); zero-dimensional quantum dots; one-dimensional nanowires; and even three-dimensional materials such as metal-organic frameworks (MOFs) and perovskites. The nonlinear optical properties of these materials enable SAs to modulate the circulating light within a laser cavity. This modulation leads to phase synchronization among longitudinal modes, producing periodic short-pulse trains in the time domain. The underlying mechanism—nonlinear absorption—is primarily driven by phenomena such as Pauli blocking or plasmonic excitation [5]. When exposed to intense light, a material may enter a "saturated" or "bleached" state. In contrast, at low light intensities, the material response is linear: the light's amplitude, phase, or polarization may change, but its frequency content remains unchanged.

Since low-dimensional materials are typically produced as thin, nanoscale flakes, they cannot be directly integrated into laser systems. Instead, they must be coupled with suitable optical structures to enable effective interaction with the light field. The approach to integration varies between solid-state and fiber lasers. In solid-state systems, the materials are often incorporated in free-space setups, allowing straightforward interaction. Fiber lasers, however, require more tailored designs to accommodate the optical fiber structure. Several common configurations are summarized in **Figure 2**. For example, candidate materials can be inserted into the hollow core of photonic crystal fibers (**Figure 2a**) [6], ensuring sufficient interaction length. Another approach is to embed the material in a polymer film (e.g., PVA or PMMA), which can then be placed between fiber connectors for transmission-type coupling (**Figure 2b**) [7] or between a connector and a reflective surface (**Figure 2c**) [8, 9]. Additionally, materials can be coated onto microfiber surfaces, where they interact with the evanescent field of the propagating light (**Figure 2d**) [10, 11].

Fiber lasers are capable of operating across a wide range of wavelengths—from visible to near- and mid-infrared—depending on the chosen saturable absorber and gain materials (e.g., rare-earth ions) [5, 12–19]. This flexibility makes it

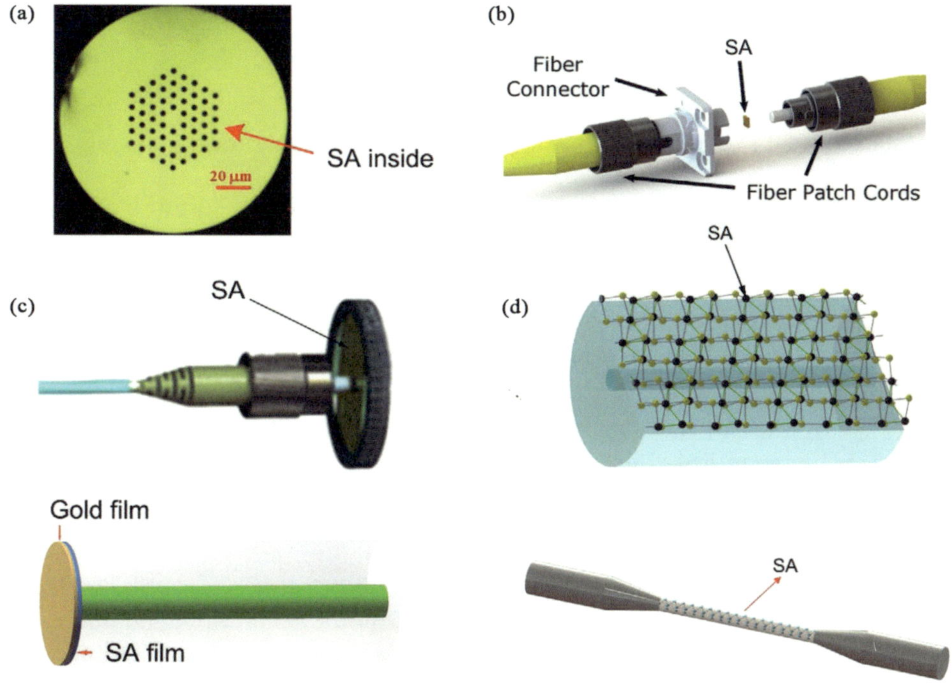

Figure 2.
(a) SA into hollow photonic crystal fibers (PCFs) [6]. (b) SA sandwiched between two fiber connectors [7]. (c) Reflective structure: above is a saturable absorber mirror [8]. Below is an optical fiber end coating [9]. (d) Microfibers: above is the D-type fibers [10]. Below are the tapered fibers [11].

possible to design short-pulse lasers for a broad spectrum of applications. Mode-locked lasers generate ultrashort pulses (on the order of femtoseconds to picoseconds) and have broad spectral bandwidths. To achieve mode-locking, the laser cavity must support a pulse shaping mechanism, where pulsed operation benefits from a higher net gain than continuous-wave (CW) operation. The saturable absorber plays a crucial role in narrowing the pulse during each pass, eventually reaching a balance between pulse compression (aided by nonlinear effects like the Kerr effect) and pulse broadening (often caused by dispersion), resulting in stable pulse formation.

Q-switched lasers, while also producing pulsed output, differ significantly in their characteristics. They emit longer pulses—typically in the nanosecond to microsecond range—with higher pulse energies compared to mode-locked systems. Although the setup for Q-switching often resembles that used in mode-locking, the outcome depends on the specific properties of both the laser cavity and the saturable absorber. Whether a particular SA-laser configuration supports Q-switching or mode-locking has been the subject of detailed studies, such as the work discussed in Ref. [20].

In this chapter, we begin by outlining the key physical mechanisms behind nonlinear absorption and exploring the nonlinear optical behaviors of various functional materials. We then focus on the design and operation of short-pulse fiber lasers that incorporate these materials as saturable absorbers, highlighting their performance and potential applications.

2. Optical nonlinearity

2.1 Kerr effect and saturable absorption

When exposed to intense electromagnetic fields, certain materials no longer respond to light in a purely linear manner. This nonlinear behavior is key to advancing photonic and optoelectronic technologies. On a macroscopic level, optical nonlinearity can be described by expressing the material's polarization, $P(t)$, as a power series expansion in terms of the incident electric field $E(t)$ [13]:

$$P(t)=\varepsilon_0\left(\chi^{(1)}E(t)+\chi^{(2)}E(t)+\chi^{(3)}E(t)+\ldots\right) \tag{1}$$

Here, $\chi^{(n)}$ represents the n-th order nonlinear susceptibility. The first term, $\chi^{(1)}$, accounts for linear optical effects such as absorption and refraction, where the response is directly proportional to the electric field. However, as the intensity of $E(t)$ increases or if the material has a significant nonlinear susceptibility, higher-order terms become prominent. The second-order susceptibility, $\chi^{(2)}$, governs processes such as second-harmonic generation, while $\chi^{(3)}$, the third-order susceptibility, leads to phenomena such as four-wave mixing, third-harmonic generation, and nonlinear refraction. Notably, the real part of $\chi^{(3)}$ corresponds to the Kerr effect (nonlinear change in refractive index), whereas the imaginary part describes nonlinear absorption phenomena, including saturable absorption.

In fiber optics, the refractive index n can vary with light intensity and is typically modeled as:

$$n=n_0+n_2I \tag{2}$$

where n_0 is the linear refractive index, n_2 is the nonlinear index coefficient, and I is the light intensity. A positive n_2 means the refractive index increases with intensity, and a negative value implies the opposite. In contrast, for linear optics, n remains constant, independent of intensity.

Saturable absorption, unlike the Kerr effect, is an inelastic process—energy is exchanged between the light and the material. It is primarily associated with the imaginary part of $\chi^{(3)}$. While saturable absorption is one form of nonlinear absorption, others include two-photon absorption (TPA) and free-carrier absorption (FCA), particularly in semiconductors. The mechanisms behind saturable absorption vary [5, 21]: in 2D materials, it is often attributed to Pauli blocking, while in metallic nanostructures, plasmonic excitations can enhance nonlinear absorption.

Under strong illumination, a material may enter a "saturated" or "bleached" state, where its ability to absorb light diminishes. The light absorption behavior in such cases is modeled by:

$$\alpha=\alpha_{ns}+\frac{\alpha_s}{1+I/I_{sat}} \tag{3}$$

In this equation, α_{ns} and α_s are the non-saturable and saturable components of the absorption coefficient, respectively; I is the incident light intensity, and I_{sat} is the saturation intensity. At low intensities, absorption is high ($\alpha_{ns}+\alpha_s$), while at high intensities, it drops to α_{ns}.

When the incident light is extremely strong, TPA becomes relevant, and the model is extended to:

$$\alpha = \alpha_{ns} + \frac{\alpha_s}{1 + I / I_{sat}} + \beta I \tag{4}$$

where β is the TPA coefficient. The corresponding light transmission, T, can be approximated as [13]:

$$\begin{aligned} T = \exp(-\alpha) &= \exp\left(-\alpha_{ns} - \frac{\alpha_s}{1 + I / I_{sat}} - \beta I\right) \\ &\approx \exp(-\alpha_{ns})\left(1 - \frac{\alpha_s}{1 + I / I_{sat}} - \beta I\right) = T_0 - \frac{T_0 \alpha_s}{1 + I / I_{sat}} - T_0 \beta I \end{aligned} \tag{5}$$

Here, $T_0 = \exp(-\alpha_{ns})$ represents the low-intensity transmission. T_0 account for material thickness, the absorption coefficient is normalized as $\alpha_0 = \alpha/L$, with L being the sample thickness. Absorbance A is related to transmission by $T = 10^{-A}$, and $A = 0.434\alpha$.

However, the basic saturable absorption model (Eqs. (3)–(5)) often fails to accurately predict experimental behavior, especially when the intensity range required for transition from unsaturated to saturated absorption is very broad. To better match experimental data, an alternative empirical model is widely adopted:

$$T = T_0 - \Delta T exp(-I / I_{sat}) - T_0 \beta I \tag{6}$$

This formulation allows for a more realistic description of the transition behavior, especially in practical photonic applications where intensity thresholds vary significantly.

2.2 Measurement of Kerr effect and saturable absorption

To characterize nonlinear optical properties such as saturable absorption and the Kerr effect, two widely used experimental methods are the *Z-scan technique* and *double-balanced detection*. These methods enable the measurement of how a material's optical properties change in response to varying light intensities.

A standard Z-scan setup is illustrated in **Figure 3a**. In this configuration, the sample is mounted on a translation stage that moves along the Z-axis (the direction of light propagation). A lens focuses the laser beam onto the sample, and two types of measurements can be performed depending on the position and state of an aperture placed before the second detector [22–24]: *open-aperture (OA)* and *closed-aperture (CA)*.

- In the *open-aperture Z-scan* mode, the aperture is removed or fully open, allowing all transmitted light to reach the detector. This mode is primarily used to evaluate *saturable absorption*, as it directly measures how the sample's transmission varies with light intensity.

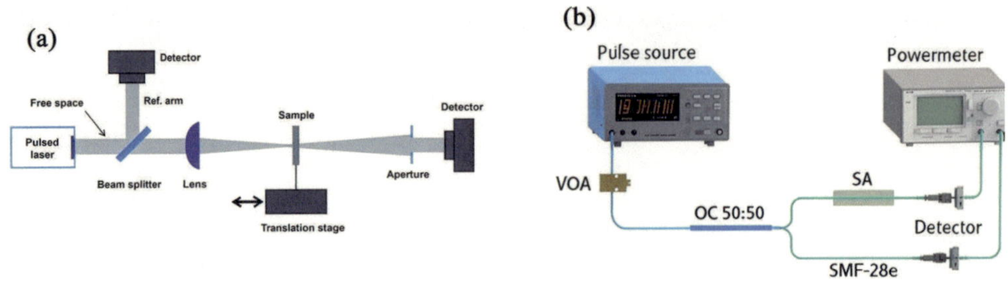

Figure 3.
Experimental setup for nonlinear absorption measurement: (a) Z-scan test device and (b) experimental setup of nonlinear absorption measurement.

- In the *closed-aperture* mode, the aperture partially blocks the beam, making the measurement sensitive not only to transmitted intensity but also to changes in the beam's divergence. This enables detection of *nonlinear refraction*, such as the *Kerr effect*, which arises when the refractive index of the sample changes with light intensity. From this, the *real part of the third-order susceptibility* $\chi^{(3)}$ and the *nonlinear refractive index* n_2 can be extracted.

As an alternative to the Z-scan method, double-balanced detection offers a simpler, fiber-based approach for measuring intensity-dependent absorption. The schematic for this setup is shown in **Figure 3b**. In this method, all optical paths are confined within optical fibers, which makes the setup compact and stable. A pulsed laser source (typically a femtosecond laser) sends light through a variable optical attenuator and into a fiber coupler, which splits the beam into two paths:

- One part serves as a *reference* and is directly measured by a detector.
- The other passes through the *sample material* before reaching the second detector.

By adjusting the input intensity via the attenuator and comparing the signals from both detectors, the material's *saturable absorption behavior* can be quantified. However, since light remains confined within the fiber and no spatial beam information is available, this method does *not* provide data on refractive index changes or beam profile evolution. Therefore, it is *limited to measuring nonlinear absorption*, unlike the more comprehensive Z-scan technique.

3. SAs based on emerging materials

The development of SA technologies has closely paralleled the evolution of laser systems. As early as 1964, passive Q-switching was demonstrated using a reversibly bleachable dye [25] and colored glass filters [26] in ruby lasers—just 4 years after Maiman's pioneering realization of laser action [1]. Despite these early breakthroughs, generating stable mode-locked pulses using passive SAs in fiber systems remained difficult until the introduction of the *SESAM* in the early 1990s [27]. SESAMs enabled compact and stable ultrafast lasers but suffer from several limitations, including narrow operational bandwidths, high fabrication costs, and picosecond-limited recovery times unless advanced and costly post-processing techniques are used [2].

These limitations have sparked significant interest in nanomaterials as next-generation SAs. Owing to their reduced dimensionality, nanomaterials often exhibit quantum confinement effects, unique electronic structures, and enhanced light–matter interactions. Broadly, they can be categorized based on dimensionality into *0D* (e.g., quantum dots), *1D* (e.g., nanowires), and *2D* materials (e.g., graphene and its derivatives).

The application of *0D materials* as SAs began in 1997 with the use of CdSe quantum dots [28]. Since then, a wide range of quantum dots—such as CdTe/CdS, $CsPbBr_3$, and PbS—have been explored for their nonlinear absorption properties, often characterized using techniques like Z-scan. These materials have enabled ultrafast pulse generation in the 1.0–1.5 μm wavelength range. Additionally, *plasmonic 0D nanomaterials*, [29, 30] including both noble and non-noble metal-based nanoparticles, have shown promise as fiber-compatible SAs. Examples include metal oxides (e.g., ITO, TCO, WO_x, In: CdO) and chalcogenides (e.g., Cu_2Se, Cu_2Te), extending the operational range up to 3.6 μm [14].

1D materials, such as carbon nanotubes, metallic and semiconductor nanowires (e.g., gold, silver, Bi_2Te_3, InP, Sb_2Se_3, GaAs, α-Bi_4Br_4) [31], also exhibit strong nonlinear optical responses. Interestingly, some materials initially considered 0D can form 1D structures under certain conditions and exhibit enhanced saturable absorption properties in this form.

The discovery of *graphene* marked a turning point in *2D material* research, revealing a new class of layered materials with exceptional optoelectronic characteristics [32]. Following graphene, several 2D materials have been extensively investigated, including *TIs [33–35]*, *TMDs [36–38]*, *BPs [39]*, and *MXenes* (e.g., $Ti_3C_2T_x$) [39]. These materials offer broad spectral operation, ultrafast recovery times, and strong third-order nonlinearities, making them attractive for pulse generation applications.

Many of these materials have been quantitatively assessed for their nonlinear optical parameters. For instance:

- *TMDs nanosheet dispersions* exhibit real parts of $\chi^{(3)}$ ranging from 3×10^{-9} to 7.75×10^{-9} esu, and nonlinear refractive indices n_2 from 9.32×10^{-7} to 6.49×10^{-7} cm^2/W.

- *TMDs thin films* show $Re\chi^{(3)}$ values between -1.26×10^{-10} and 4.82×10^{-9} esu, with n_2 values ranging from $-3.36 \times 10^{-12} \sim 1.28 \times 10^{-10}$ cm^2/W.

- *TI Bi_2Te_3 dispersions* typically exhibit $Re\chi^{(3)}$ around 10^{-7} esu and n_2 2.11×10^{-8} cm^2/W.

- *BPs dispersions* have reported $Re\chi^{(3)}$ values between 2.56×10^{-8} and 4.18×10^{-8} esu, and n_2 between 2.27×10^{-5} and 3.71×10^{-5} cm^2/W.

(For a more comprehensive overview of these values, please refer to Ref. [4]).

In the following section, we will examine representative examples of saturable absorbers across different material classes, demonstrating their effectiveness in generating ultrafast laser pulses.

4. Theoretical design criteria for saturable absorbers

Pulsed lasers typically operate through one of two mechanisms: *Q-switching* or *mode-locking*. Both can be achieved using either active modulation—which involves external electronic or mechanical control *via* electro-optic or acousto-optic modulators—or passive modulation, which relies on intrinsic nonlinear optical (NLO) processes such as saturable absorption, Kerr nonlinearity, and nonlinear polarization rotation.

Q-switching is commonly used to generate pulses with high energy. In this method, the *quality factor* (Q) of the laser cavity—defined as the ratio of stored energy to energy lost per cycle—is modulated to control the timing of energy release. Initially, the cavity operates at a low Q value to accumulate energy while suppressing lasing. When Q is suddenly increased (e.g., by opening a "shutter" inside the cavity), the stored energy is rapidly released as a giant pulse. In *passive Q-switching*, this shutter is effectively replaced by a saturable absorber, which modulates Q based on light intensity without external control.

A theoretical framework for understanding cw Q-switching was developed by Kärtner et al. in 1995 [40], considering the interplay between laser cavity configuration, gain material, and SA properties. When designing pulsed lasers—particularly those using SAs—several fundamental parameters must be taken into account. A critical factor is the saturation energy of both the gain medium and the absorber, which can be expressed as:

$$E_{sat,\mathrm{L/A}} = F_{sat,\mathrm{L/A}} \times A_{eff,\mathrm{L/A}} \tag{7}$$

Here, F_{sat} is the saturation fluence, and A_{eff} is the effective mode area for the gain material (L) or absorber (A). The saturation fluence of the gain medium is given by:

$$F_{sat,\mathrm{L}} = h\upsilon / m\sigma_{\mathrm{L}} \tag{8}$$

where:

h is Planck's constant,

v is the photon frequency,

σ_{L} is the emission cross-section of the gain material,

and m is a cavity parameter, with $m = 1$ for a ring cavity and $m = 2$ for a linear cavity.

The effective mode area is typically calculated using the beam radius ω_{eff} as:

$$\mathrm{A}_{eff,\mathrm{L/A}} = \pi\omega_{eff,\mathrm{L}} / \mathrm{A}^2 \tag{9}$$

The saturation fluence of the absorber, $F_{\mathrm{sat,A}}$, can be experimentally measured, often using the Z-scan technique, a widely adopted method for characterizing nonlinear optical materials.

To ensure stable cw Q-switching, a condition involving saturation fluence, modulation depth ΔR, and recovery time τ_A must be satisfied:

$$F_{sat,\mathrm{L}} / \Delta R\tau_A \leq \frac{2hc\omega_{eff,\mathrm{L}}^2}{m\lambda\sigma_{em,\mathrm{L}}\omega_{eff,\mathrm{A}}^2 T_R} \tag{10}$$

This inequality highlights the interaction between the nonlinear optical properties of the absorber (left-hand side) and the characteristics of the laser cavity and gain medium (right-hand side). For effective Q-switching, it is generally advantageous to use gain materials with long upper-state lifetimes and short cavity lengths. SAs with moderate to high modulation depth and longer recovery times tend to perform better in this regime. However, increasing modulation depth too much can introduce excessive intracavity losses, so a careful balance between gain and loss must be maintained for optimal laser performance.

In contrast, *mode-locking* is a preferred approach when the goal is to produce ultrashort pulses—typically on the order of picoseconds or femtoseconds—which are much shorter than the round-trip time of the cavity. Theoretical studies on mode-locking began in the 1970s [41], and although the foundational principles are established, the exact dynamics are still an active area of research. To achieve mode-locked pulse generation, certain key conditions must be met. A fundamental requirement is that the intracavity pulse energy E_P must exceed a threshold value $E_{P,C}$, which depends on the properties of both the SA and the laser gain medium. This relationship is described as:

$$E_P^2 > E_{\mathrm{P,c}}^2 = E_{sat,\mathrm{L}} E_{sat,\mathrm{A}} \Delta R \tag{11}$$

Here, $E_{\mathrm{sat,L}}$ and $E_{\mathrm{sat,A}}$ represent the saturation energies of the laser gain material and the saturable absorber, respectively, and ΔR is the modulation depth of the SA.

This condition can also be reformulated using saturation fluences and cavity parameters:

$$F_{sat,\mathrm{A}} \Delta R < \frac{(PT_R)^2}{F_{sat,\mathrm{L}} \times A_{sat,\mathrm{L}} \times A_{sat,\mathrm{A}}} = \frac{(PT_R)^2 \times m\sigma_{em,\mathrm{L}} \lambda}{hc \times \pi\omega_{eff,\mathrm{L}}^2 \times \pi\omega_{eff,\mathrm{A}}^2} \tag{12}$$

In this expression:

- F_{sat} is the saturation fluence,
- A_{sat} is the effective mode area,
- T_R is the cavity round-trip time,
- $\sigma_{\mathrm{em,L}}$ is the emission cross-section of the gain medium,
- λ is the laser wavelength,
- h is Planck's constant, and
- c is the speed of light.

On the left side of the inequality are parameters related to the nonlinear optical properties of the saturable absorber. On the right are terms determined by the gain material and cavity configuration. For stable mode-locking, both must be carefully considered and balanced during design.

Compared to Q-switching, mode-locking imposes more stringent requirements on the saturable absorber. Specifically:

- The SA should have a low saturation fluence and a moderate modulation depth.
- Its recovery time must be shorter than the cavity round-trip time, especially important for compact lasers with short cavities (typically a few centimeters or less).

If the recovery time is too long, the SA cannot respond quickly enough to support ultrashort pulse formation. Additionally, while a higher modulation depth helps initiate mode-locking, excessive depth can introduce losses. Therefore, a balanced design is crucial.

The relationship between the pulse width τP and the modulation depth ΔR can be approximated as:

$$\tau_P \propto (2/\Delta R)^r \tag{13}$$

where r is a positive exponent. This indicates that a larger modulation depth can support the formation of shorter pulses, although practical considerations often require a compromise among modulation depth, loss, and pulse stability.

When using low-dimensional materials (e.g., quantum dots, nanowires, and 2D materials like graphene) as SAs in fiber lasers, there are several key design considerations. Various all-fiber laser configurations (as shown in **Figure 2**) integrate these materials for pulse generation. Among the most critical SA parameters are [2, 42, 43]:

1. *Non-saturable loss*: This arises from surface scattering and intrinsic or extrinsic linear absorption. High non-saturable loss reduces laser efficiency and increases thermal load, potentially causing device failure. While some loss is unavoidable, material engineering should aim to minimize it.

2. *Damage threshold*: SAs must withstand high peak powers without degradation. Materials with low thermal or chemical stability are unsuitable. Ideal SAs should have high damage thresholds and remain stable under ambient conditions.

3. *Operating wavelength and bandwidth*: These depend on the material's absorption spectrum. While broadband SAs are attractive, achieving wide operational ranges often involves trade-off with other properties, such as reduced modulation depth or increased loss.

4. *Cost and integration compatibility*: For widespread application, SAs must be cost-effective and compatible with existing fiber-optic platforms. High fabrication costs or difficult integration processes can hinder commercial adoption.

In summary, the theoretical design of SAs for Q-switching or mode-locking requires a careful balance of optical, thermal, and mechanical properties. Selecting appropriate materials and optimizing their integration with laser cavities are essential steps toward high-performance pulsed laser systems.

5. Fiber lasers based on emerging saturable absorbers

Pulsed fiber lasers—whether Q-switched or mode-locked—depend critically on an effective SA. Ideal SAs exhibit low non-saturable losses, strong nonlinear response, fast recovery, and are easy and inexpensive to produce. In this section, we explore how SAs based on 0D, 1D, and 2D nanomaterials shape pulse generation in fiber lasers, using representative examples rather than exhaustive surveys.

5.1 0D SAs for fiber lasers

QDs, such as those made from MoS_2, graphene, BP, carbon, and other nanomaterials, have been widely investigated for their application in fiber lasers. These 0D materials exhibit strong nonlinear optical responses and ultrafast carrier dynamics, making them suitable for pulse generation in both Q-switched and mode-locked regimes.

Among them, heavily doped semiconductors, particularly doped oxide materials, have gained attention due to their tunable optical properties and compatibility with fabrication techniques. One such example is aluminum-doped zinc oxide (AZO) nanocrystals, which exhibit strong nonlinear absorption across a broad infrared (IR) spectrum (800–3000 nm) [3]. Characterized using Z-scan and pump-probe techniques, AZO shows a nonlinear absorption coefficient as high as -1.90×10^3 cm/GW and an ultrafast optical response of less than 350 fs. When embedded into a polyvinyl alcohol (PVA) film, the resulting AZO/PVA SA can be integrated into a fiber laser ring cavity.

Experiments using this AZO-SA demonstrated its effectiveness in both Yb-doped (YDF) and Er-doped fiber (EDF) laser systems [3]. As shown in **Figure 4a** and **b**, passively mode-locked lasers were constructed with ring cavity configurations. The resulting spectra (**Figure 4c** and **d**) showed stable mode-locked pulses centered at 1061.7 and 1533.8 nm, respectively. The radio-frequency (RF) spectra confirmed strong mode-locking stability. A pulse duration of 740 ps was observed, with the shortest achievable pulse calculated to be ~1.39 ps. In the case of the Yb-doped laser operating in the all-normal dispersion regime, longer pulses were expected but could be further compressed using additional dispersion management. In the Er-doped setup, the presence of fixed sidebands in the soliton spectrum (**Figure 4d**) confirmed operation in the soliton regime with anomalous dispersion. Impressively, femtosecond pulses of 622 fs (with a minimum of 380 fs) were achieved.

AZO nanocrystals also performed well in Tm-doped fiber (TDF, **Figure 5a**) and Ho^{3+}/Pr^{3+}-codoped ZBLAN fiber lasers [3], enabling passive Q-switching at longer

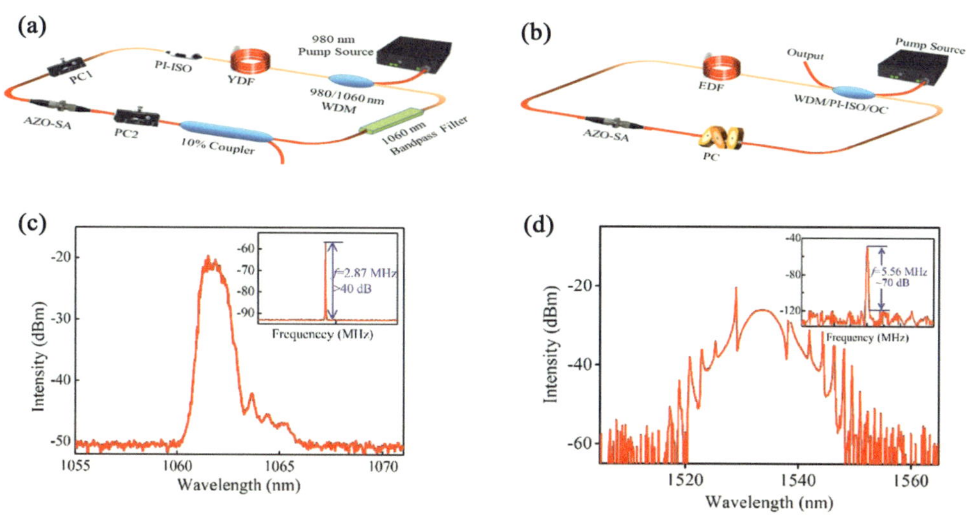

Figure 4.
(a) Mode-locking operation in YDF laser based on AZO SA. (b) Mode-locking operation in EDF laser based on AZO SA. (c) Mode-locking output pulse spectrum in YDF laser, centered at 1061.7 nm. The inset is the fundamental frequency of a typical RF spectrum. (d) Mode-locking output pulse spectrum in EDF laser, centered at 1533.8 nm. The inset is the fundamental frequency of a typical RF spectrum.

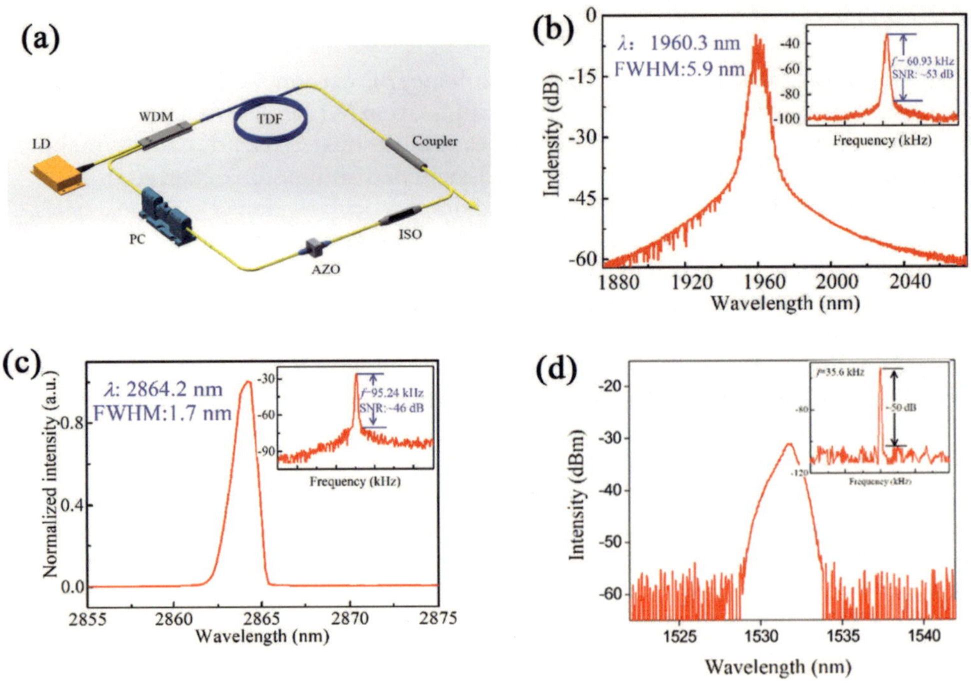

Figure 5.
(a) Q-switching operation in TDF laser based on AZO SA. (b) Q-switching optical spectrum in TDF laser. The inset is an RF spectrum at the fundamental frequency. (c) Q-switching optical spectrum in Ho/Pr codoped ZBLAN fiber laser. The inset is an RF spectrum at the fundamental frequency. (d) Q-switching optical spectra in EDF laser based on $Cu_{1.3}Se$ SA, with a center wavelength of 1533.8 nm.

wavelengths. As illustrated in **Figure 5b–c**, stable Q-switched pulses were generated at 1960.3 nm (TDF, **Figure 5b**) and 2864.2 nm (Ho^{3+}/Pr^{3+}-codoped ZBLAN fiber, **Figure 5c**). Theoretical pulse durations were estimated at 1.55 and 1.16 μs for each wavelength, respectively.

Another promising 0D material is $Cu_{1.3}Se$ nanocrystals [29], which are easily synthesized *via* solution processing. These materials exhibit both ultrafast recovery times (360–520 fs) and broadband nonlinear absorption from 800 to 3000 nm, with a high nonlinear coefficient (-1.4×10^3 cm/GW). $Cu_{1.3}Se$ SAs, similarly embedded in PVA films, were used in erbium-doped fiber lasers. At a pump power of 12 mW, stable Q-switching was observed, with the output spectrum centered at 1533.8 nm and a full-width at half maximum (FWHM) of 3.33 nm (**Figure 5d**). RF analysis revealed a signal-to-noise ratio exceeding 50 dB, indicating strong pulse stability.

Notably, the $Cu_{1.3}Se$-based Q-switching behavior differed from the mode-locking dynamics observed with AZO-based SAs, reflecting distinct material properties. In addition to EDF systems, $Cu_{1.3}Se$ was successfully applied in TDF and Ho^{3+}/Pr^{3+}-codoped ZBLAN fiber lasers, producing stable Q-switched pulses at 1960.3 and 2864.6 nm, further demonstrating its versatility as a saturable absorber.

5.2 1D SAs for fiber lasers

Carbon nanotubes (CNTs) represent one of the most widely studied one-dimensional (1D) materials in ultrafast photonics. In this section, we highlight the use of CNT-based SAs in shaping pulsed outputs from fiber lasers. Since the first demonstration of single-walled carbon nanotube (SWCNT) SAs in 2003 [44], these materials have attracted widespread attention due to their simple fabrication process, low cost, and broad optical response.

One of the key advantages of CNTs is their sub-picosecond recovery time, which makes them particularly effective for ultrafast pulse generation. Additionally, their natural variation in diameter during synthesis leads to broadband absorption, enhancing their applicability across various wavelength regimes.

CNT-based SAs have been successfully used to achieve mode-locking in Yb-, Er-, and Tm:Ho-doped fiber lasers, operating at wavelengths of approximately 1.05, 1.56, and 1.99 μm, respectively [45]. In the experimental setup illustrated in **Figure 6a**, a linear cavity design was adopted. The setup included a pump laser, wavelength-selective pump coupler, fiber output coupler (~10%), a highly reflective dielectric mirror, and the SWCNT absorber mirror. A polarization controller was included to fine-tune performance.

Notably, Er- and Tm:Ho-doped lasers operated directly in the soliton regime without the need for additional dispersion compensation. In contrast, Yb-doped fiber lasers, which typically exhibit normal dispersion, require a grating pair to manage dispersion and enable stable mode-locking.

The lasers were pumped using fiber-coupled 976 nm diodes for the Yb and Er systems, while the Tm:Ho laser was pumped by an Er-fiber laser emitting at 1.56 μm. The CNT SA mirror was assembled *via* fiber butt-coupling and typically experienced oversaturation from the intracavity pulse fluence. The results demonstrated effective ultrafast pulse generation across the three systems: The Yb-fiber laser operated at 1050 nm with a spectral bandwidth of 2.7 nm. The Er-doped fiber laser produced 0.44 ps pulses at 1563 nm, with a spectral width of 6.0 nm and a time-bandwidth

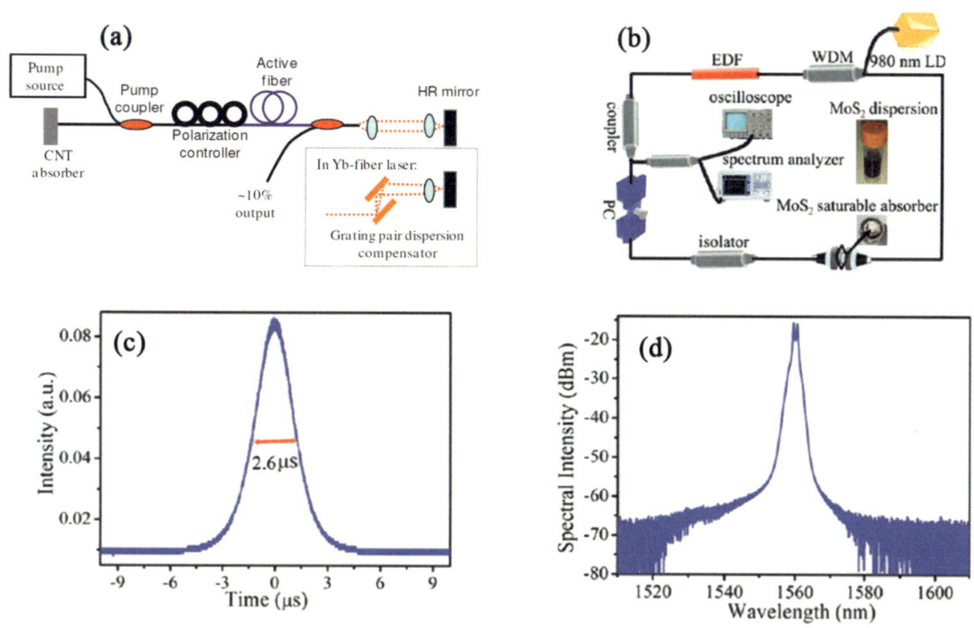

Figure 6.
(a) Schematic of the experimental fiber laser setup based on CNT-SA. (b) Schematic diagram of a Q-switched fiber laser based on MoS_2 nanoflowers-SA. (c) Single Q-switching pulse profile. (d) output spectrum of the Q-switching pulse.

product of 0.32, indicating near-transform-limited performance. The Tm:Ho fiber laser delivered pulses of approximately 1.0 ps centered at 1991 nm.

These findings confirm the suitability of CNT-based SAs for generating ultrashort pulses across a wide spectral range, offering a robust and cost-effective solution for diverse fiber laser systems.

5.3 2D SAs for fiber lasers

The discovery of graphene has paved the way for extensive research into other 2D materials with exceptional optical and electronic properties. These materials have enabled the development of a range of photonic devices, including SAs, optical modulators, photodetectors, and optical switches. Among them, molybdenum disulfide (MoS_2) has attracted considerable attention due to its promising NLO behavior [42, 46].

Studies have shown that the NLO response of MoS_2 is highly dependent on its thickness and lateral size. For example, MoS_2 nanosheets thicker than 8 nm typically exhibit saturable absorption (SA), whereas thinner sheets around 4 nm may show reverse saturable absorption (RSA) [47]. Similarly, lateral size plays a crucial role: sheets larger than 100 nm often support SA, while those under 50 nm tend to exhibit RSA [43]. These findings suggest that both geometry and morphology significantly influence the nonlinear performance of 2D materials.

In one study, MoS_2 nanoflowers were embedded in a PVA matrix to create a composite film [43]. This MoS_2/PVA saturable absorber was used in a passively Q-switched EDF laser (**Figure 6b**), operating at a wavelength of 1.56 μm. The system demonstrated a high modulation depth (51.8%) and saturable intensity (275.5 GW/cm^2). The laser cavity consisted of a 2 m long Er-doped gain fiber and a 10 m single-mode fiber (SMF-28e), with a 980 nm pump light injected via a wavelength division multiplexer (WDM). A 90:10 fiber coupler was used for output coupling, while a polarization controller (PC) and isolator ensured stable unidirectional operation.

The MoS_2/PVA film was integrated between two fiber connectors. Stable Q-switched pulses were generated at a low pump threshold of 16.1 mW. The pulses exhibited a repetition rate of 45.4 kHz and an average output power of 2.105 mW, with a Gaussian-like pulse profile and a duration of 2.6 μs (**Figure 6c**). The optical spectrum is centered at 1560 nm (**Figure 6d**). The corresponding radio-frequency (RF) spectrum revealed a signal-to-noise ratio (SNR) of ~52.5 dB, indicating excellent stability. No additional noise peaks were observed within a 300 kHz frequency span, confirming the laser's consistent Q-switched operation. These results highlight the strong potential of MoS_2 nanoflowers as effective SAs in fiber laser applications.

Another noteworthy 2D material is bismuth oxyselenide (Bi_2O_2Se) [35], a topological insulator with high electron mobility (~3000 cm^2/V s at 1.9 K), excellent environmental stability, and compatibility with scalable fabrication methods. Its electronic structure, primarily composed of p orbitals from Bi and Se atoms, facilitates strong spin-orbit interactions and supports nonlinear optical responses potentially enhanced by topological surface states.

Recent work by X.L. Tian et al. demonstrated the use of Bi_2O_2Se-based saturable absorbers in fiber lasers operating at mid-infrared wavelengths (~3 μm) [35]. The absorber showed an ultra-broadband nonlinear response ranging from 0.8 to 5.0 μm. The experimental setup employed a double-cladding Ho^{3+}/Pr^{3+} codoped ZBLAN fiber as the gain medium, pumped by two diode lasers (~1150 nm) combined through a polarized beam splitter (PBS). **Figure 7a** illustrates the compact laser configuration.

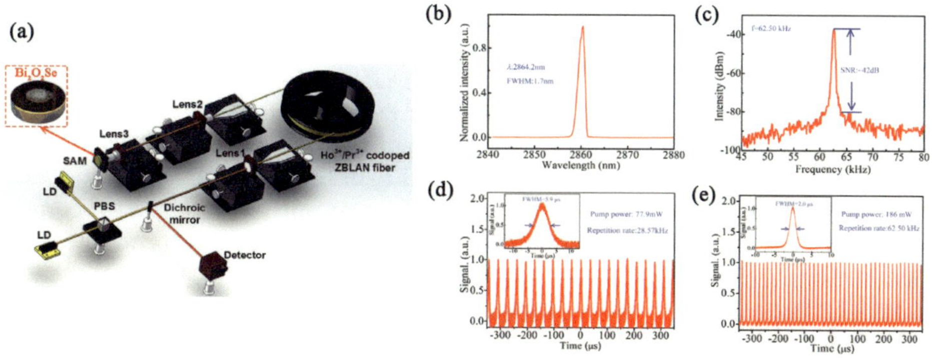

Figure 7.
(a) Schematic of the fiber laser setup based on Bi_2O_2Se SA, (b) Typical Q-switching optical spectrum, (c) The RF spectrum at fundamental frequency. Typical Q-switched pulse trains at the pump power of: (d) 77.9 mW and (e) 186 mW.

Stable Q-switched pulses were first observed when the launched pump power reached approximately 77.9 mW and remained stable up to 186 mW. A typical output spectrum centered around 2864.2 nm with a full-width at half maximum (FWHM) of 1.7 nm (**Figure 7b**). The corresponding RF spectrum showed an SNR of over 42 dB (**Figure 7c**), confirming the stability of the Q-switching regime. At a pump power of 77.9 mW, the pulse repetition rate was 28.57 kHz, with a pulse width of 5.9 μs (**Figure 7d**). As the pump power increased, the repetition rate rose nearly linearly to 62.5 kHz, while the pulse width decreased to 2.0 μs (**Figure 7e**), in line with typical Q-switching behavior. The highest output power recorded was 21.5 mW, corresponding to a pulse energy of ~0.344 μJ.

Compared to other low-dimensional materials, such as graphene, WS_2, and black phosphorus, Bi_2O_2Se offers distinct advantages in broadband optical modulation and processing flexibility, making it a promising candidate for next-generation photonic devices across a wide spectral range.

6. Conclusion and outlook

In this chapter, we have explored the pivotal role of saturable absorbers in shaping the performance of pulsed fiber lasers. Beginning with a discussion of the fundamental mechanisms of nonlinear optics—specifically, saturable absorption and the Kerr effect—we established the foundation for understanding how functional materials interact with light in intensity-dependent ways. Using both theoretical models and practical measurement techniques, such as the Z-scan and double-balanced detection, we highlighted how these nonlinearities can be quantified and exploited.

Low-dimensional materials, including 0D quantum dots, 1D carbon nanotubes, and 2D materials such as MoS_2, graphene, and Bi_2O_2Se, have proven particularly effective as SAs due to their strong optical nonlinearity, ultrafast carrier dynamics, and structural tunability. Demonstrated in a range of fiber laser systems—from Yb- and Er-doped lasers to mid-infrared Tm- and Ho-doped systems—these materials support both Q-switching and mode-locking across a broad spectral range. The successful integration of these SAs has led to significant improvements in pulse duration, repetition rate, and stability.

Looking ahead, several challenges and opportunities remain. First, optimizing the trade-offs between modulation depth, recovery time, and non-saturable loss is crucial for next-generation laser design. Second, material stability under high-intensity and long-term operation must be further improved to ensure reliability in practical applications. Finally, the continued development of scalable, low-cost fabrication techniques will be key to the commercial viability of SA-based photonic devices.

In conclusion, functional nanomaterials offer a promising path toward compact, energy-efficient, and high-speed pulsed laser systems. As synthesis and integration techniques continue to mature, we expect these materials to play a central role in expanding the capabilities of ultrafast photonics across scientific, industrial, and medical domains.

Acknowledgements

This work was supported by the National Natural Science Foundation of China (Grant No. 12204435), the Natural Science Foundation of Zhejiang Province (Grant No. LQ22A040011), and the Municipal Government of Quzhou (Grant Nos. 2023D012, 2023D013, and 2024D020).

Conflict of interest

The authors declare no conflict of interest.

Author details

Xiangling Tian[1*], Wenhao Wei[1], Peng Bi[1] and Rongfei Wei[2]

1 Yangtze Delta Region Institute (Quzhou), University of Electronic Science and Technology of China, Quzhou, Zhejiang, China

2 Department of Physics, Zhejiang Normal University, Jinhua, Zhejiang, China

*Address all correspondence to: xianglingt@sina.com

References

[1] Maiman TH. Stimulated optical radiation in ruby. Nature. 1960;**187**: 493-494. DOI: 10.1038/187493a0

[2] Keller U. Recent developments in compact ultrafast lasers. Nature. 2003;**424**:831-838. DOI: 10.1038/nature01938

[3] Tian X, Luo H, Wei R, Liu M, Yang Z, Luo Z, et al. Ultrafast and broadband optical nonlinearity in aluminum doped zinc oxide colloidal nanocrystals. Nanoscale. 2019;**11**:13988-13995. DOI: 10.1039/C9NR04337C

[4] Wu K, Chen B, Zhang X, Zhang S, Guo C, Li C, et al. High-performance mode-locked and Q-switched fiber lasers based on novel 2D materials of topological insulators, transition metal dichalcogenides and black phosphorus: Review and perspective (invited). Optics Communications. 2018;**406**:214-229. DOI: 10.1016/j.optcom.2017.02.024

[5] Zhang Y, Lu D, Yu H, Zhang H. Low-dimensional saturable absorbers in the visible spectral region. Advanced Optical Materials. 2019;7:1800886. DOI: 10.1002/adom.201800886

[6] Lin Y-H, Yang C-Y, Liou J-H, Yu C-P, Lin G-R. Using graphene nano-particle embedded in photonic crystal fiber for evanescent wave mode-locking of fiber laser. Optics Express. 2013;**21**:16763-16776. DOI: 10.1364/OE.21.016763

[7] Woodward RI, Howe RCT, Runcorn TH, Hu G, Torrisi F, Kelleher EJR, et al. Wideband saturable absorption in few-layer molybdenum diselenide ($MoSe_2$) for Q-switching Yb-, Er- and Tm-doped fiber lasers. Optics Express. 2015;**23**:20051-20061. DOI: 10.1364/OE.23.020051

[8] Liu W, Zhu Y-N, Liu M, Wen B, Fang S, Teng H, et al. Optical properties and applications for MoS_2-Sb_2Te_3-MoS_2 heterostructure materials. Photonics Research. 2018;**6**:220-227. DOI: 10.1364/PRJ.6.000220

[9] Chen H, Chen Y, Yin J, Zhang X, Guo T, Yan P. High-damage-resistant tungsten disulfide saturable absorber mirror for passively Q-switched fiber laser. Optics Express. 2016;**24**:16287-16296. DOI: 10.1364/OE.24.016287

[10] Cui Y, Lu F, Liu X. Nonlinear saturable and polarization-induced absorption of rhenium disulfide. Scientific Reports. 2017;7:40080. DOI: 10.1038/srep40080

[11] Li Y, He Y, Cai Y, Chen S, Liu J, Chen Y, et al. Black phosphorus: Broadband nonlinear optical absorption and application. Laser Physics Letters. 2018;**15**:025301

[12] Tian X, Ma Z, Qiu J, Wei R. Nd^{3+}-doped glass-ceramic fiber fabricated by drawing precursor ceramic and successive heat treatment. Ceramics International. 2022;**48**:21671-21675. DOI: 10.1016/j.ceramint.2022.04.147

[13] Tian X, Wei R, Wang L. Nonlinear optical effect and its applications of TMDs. In: Vikraman D, Kim H-S, editors. Distinctive Aspects of Molybdenum Disulfide. London, UK, Rijeka: IntechOpen; 2025. DOI: 10.5772/intechopen.1009547

[14] Zhao D, Liu Y, Qiu J, Liu X. Plasmonic saturable absorbers. Advanced Photonics Research. 2021;**2**:2100003. DOI: 10.1002/adpr.202100003

[15] Tian X, Wei R, Ma Z, Qiu J. Amplified spontaneous emission from

perovskite quantum dots inside a transparent glass. Advanced Optical Materials. 2022;**10**:2102483. DOI: 10.1002/adom.202102483

[16] Wei R, Chen Y, Wang L, Pan J, Tian X, Hu F, et al. High-temperature X-ray time-lapse imaging based on the improved scintillating performance of Na5Lu9F32:Tb^{3+} glass ceramics. Advanced Optical Materials. 2025;**13**:2402496. DOI: 10.1002/adom.202402496

[17] Wei R, Dai P, Tian X, Ma L, Yu Q, Hu F, et al. Excellent irradiation stability of Ce^{3+}-doped Na5Lu9-xGdxF32 glass-ceramics scintillator for X-ray imaging. Journal of the European Ceramic Society. 2024;**44**:2418-2426. DOI: 10.1016/j.jeurceramsoc.2023.11.010

[18] Li L, Wei R, Wang L, Tian X, Li X, Hu F, et al. Enhanced scintillating performance in Tb^{3+} doped oxyfluoride glass for high-resolution X-ray imaging. Ceramics International. 2024;**50**:53342-53350. DOI: 10.1016/j.ceramint.2024.10.184

[19] Ma L, Wei R, Yu Q, Dai P, Tian X, Hu F, et al. Improved broadband luminescence in Gd2GaSb1-xTaxO7:Cr^{3+},Yb^{3+} pyrochlore phosphors: Near-infrared spectroscopic applications and dual-mode optical thermometry. Materials Today Chemistry. 2024;**38**:102090. DOI: 10.1016/j.mtchem.2024.102090

[20] Hönninger C, Paschotta R, Morier-Genoud F, Moser M, Keller U. Q-switching stability limits of continuous-wave passive mode locking. Journal of the Optical Society of America B: Optical Physics. 1999;**16**:46-56. DOI: 10.1364/JOSAB.16.000046

[21] Kauranen M, Zayats AV. Nonlinear plasmonics. Nature Photonics. 2012;**6**:737-748. DOI: 10.1038/nphoton.2012.244

[22] Wei R, Tian X, Hu Z, Zhang H, Qiao T, He X, et al. Vertically standing layered MoS_2 nanosheets on TiO_2 nanofibers for enhanced nonlinear optical property. Optics Express. 2016;**24**:25337-25344. DOI: 10.1364/OE.24.025337

[23] Wei R, Zhang H, He X, Hu Z, Tian X, Xiao Q, et al. Versatile preparation of ultrathin MoS_2 nanosheets with reverse saturable absorption response. Optical Materials Express. 2015;**5**:1807-1814. DOI: 10.1364/OME.5.001807

[24] Wei R, Tian X, Zhang H, Hu Z, He X, Chen Z, et al. Facile synthesis of two-dimensional WS2 with reverse saturable absorption and nonlinear refraction properties in the PMMA matrix. Journal of Alloys and Compounds. 2016;**684**:224-229. DOI: 10.1016/j.jallcom.2016.05.169

[25] Soffer BH. Giant pulse laser operation by a passive, reversibly bleachable absorber. Journal of Applied Physics. 1964;**35**:2551-2551. DOI: 10.1063/1.1702903

[26] Bret G, Gires F. Giant-pulse laser and light amplifier using variable transmission coefficient glasses as light switches. Applied Physics Letters. 1964;**4**:175-176. DOI: 10.1063/1.1753924

[27] Zirngibl M, Stulz LW, Stone J, Hugi J, DiGiovanni D, Hansan PB. 1.2 ps pulses from passively mode-locked laser diode pumped Er-doped fibre ring laser. Electronics Letters. 1991;**27**:1734-1735. DOI: 10.1049/el:19911079

[28] Guerreiro P, Ten S, Borrelli N, Butty J, Jabbour G, Peyghambarian N. PbS quantum-dot doped glasses as saturable absorbers for mode locking of a Cr: Forsterite laser. Applied

Physics Letters. 1997;**71**:1595-1597. DOI: 10.1063/1.119843

[29] Wei R, Tian X, Luo H, Liu M, Yang Z, Luo Z, et al. Heavily doped semiconductor colloidal nanocrystals as ultra-broadband switches for near-infrared and mid-infrared pulse lasers. ACS Applied Materials & Interfaces. 2019;**11**:40416-40423. DOI: 10.1021/acsami.9b10949

[30] Tian X, Wei R, Yang D, Qiu J. Paradoxical combination of saturable absorption and reverse-saturable absorption in plasmon semiconductor nanocrystals. Nanoscale Advances. 2020;**2**:1676-1684. DOI: 10.1039/C9NA00694J

[31] Tan Z, Dong J, Liu Y, Luo Q, Li Z, Yun T, et al. Nonlinear optics of graphitic carbon allotropes: From 0D to 3D. Nanoscale. 2025;**17**:1171-1212. DOI: 10.1039/D4NR03467H

[32] Novoselov KS, Jiang D, Schedin F, Booth TJ, Khotkevich VV, Morozov SV, et al. Two-dimensional atomic crystals. National Academy of Sciences of the United States of America. 2005;**102**:10451-10453. DOI: 10.1073/pnas.0502848102

[33] Wei R, Tian X, Yang L, Yang D, Ma Z, Guo H, et al. Ultrafast and large optical nonlinearity of a $TiSe_2$ saturable absorber in the 2 μm wavelength region. Nanoscale. 2019;**11**:22277-22285. DOI: 10.1039/C9NR06374A

[34] Tian X, Wei R, Liu M, Zhu C, Luo Z, Wang F, et al. Ultrafast saturable absorption in TiS_2 induced by non-equilibrium electrons and the generation of a femtosecond mode-locked laser. Nanoscale. 2018;**10**:9608-9615. DOI: 10.1039/C8NR01573B

[35] Tian X, Luo H, Wei R, Zhu C, Guo Q, Yang D, et al. An ultrabroadband mid-infrared pulsed optical switch employing solution-processed bismuth oxyselenide. Advanced Materials. 2018;**30**:1801021. DOI: 10.1002/adma.201801021

[36] Tian X, Wei R, Liu S, Zhang Y, Qiu J. Photoluminescence nonuniformity from self-seeding nuclei in CVD-grown monolayer $MoSe_2$. Nanoscale. 2018;**10**:752-757. DOI: 10.1039/C7NR08662H

[37] Tian X, Wei R, Ma Z, Qiu J. Broadband nonlinear optical absorption induced by bandgap renormalization in CVD-grown monolayer $MoSe_2$. ACS Applied Materials & Interfaces. 2022;**14**:8274-8281. DOI: 10.1021/acsami.1c23053

[38] Tian X, Wei R, Guo Q, Zhao Y-J, Qiu J. Reverse saturable absorption induced by phonon-assisted anti-stokes processes. Advanced Materials. 2018;**30**:1801638. DOI: 10.1002/adma.201801638

[39] Guo B, Xiao Q-L, Wang S-H, Zhang H. 2D layered materials: Synthesis, nonlinear optical properties, and device applications. Laser & Photonics Reviews. 2019;**13**:1800327. DOI: 10.1002/lpor.201800327

[40] Kaertner F, Brovelli L, Kopf D, Kamp M, Calasso I, Keller U. Control of solid state laser dynamics by semiconductor devices. Optical Engineering. 1995;**34**:2024-2036. DOI: 10.1117/12.204794

[41] Haus HA. Theory of mode locking with a fast saturable absorber. Journal of Applied Physics. 1975;**46**:3049-3058. DOI: 10.1063/1.321997

[42] Wei R, Zhang H, Hu Z, Qiao T, He X, Guo Q, et al. Ultra-broadband nonlinear saturable absorption

of high-yield MoS_2 nanosheets. Nanotechnology. 2016;**27**:305203. DOI: 10.1088/0957-4484/27/30/305203

[43] Wei R, Zhang H, Tian X, Qiao T, Hu Z, Chen Z, et al. MoS_2 nanoflowers as high performance saturable absorbers for an all-fiber passively Q-switched erbium-doped fiber laser. Nanoscale. 2016;**8**:7704-7710. DOI: 10.1039/C5NR08457A

[44] Set S, Yaguchi H, Tanaka Y, Jablonski M, Sakakibara Y, Rozhin A, et al. Kikuchi, Optical Fiber Communication Conference. Atlanta, Georgia, USA: IEEE; 2003. p. PD44-P1

[45] Kivistö S, Hakulinen T, Kaskela A, Aitchison B, Brown DP, Nasibulin AG, et al. Carbon nanotube films for ultrafast broadband technology. Optics Express. 2009;**17**:2358-2363. DOI: 10.1364/OE.17.002358

[46] Wei R, Qiao T, Tian X, Zhang H, He X, Hu Z, et al. Enhanced nonlinear optical response of Se-doped MoS_2 nanosheets for passively Q-switched fiber laser application. Nanotechnology. 2017;**28**:215206. DOI: 10.1088/1361-6528/aa6bc8

[47] Khan AR, Zhang L, Ishfaq K, Ikram A, Yildrim T, Liu B, et al. Optical harmonic generation in 2D materials. Advanced Functional Materials. 2022;**32**:2105259. DOI: 10.1002/adfm.202105259

Chapter 4

Advancing Q-Switched Pulse Generation in the 2 μm Wavelength Region Using Niobium Carbide-Based Saturable Absorber

Cheng Xiau San, Lip Zhan Hong and Tan Sin Jin

Abstract

Over the past decades, lasers operating in the mid-infrared (MIR) region have attracted considerable attention because of their wide-ranging applications. In particular, fiber lasers operating near the 2 μm wavelength have attracted significant interest for use in fields such as biomedical, optical communications, and remote sensing. This chapter focuses on the development of fiber lasers operating in the 2 μm region, with an emphasis on pulsed laser operation in the near-MIR region using MXene-based thin-film saturable absorbers (SAs). In this study, niobium carbide-based (Nb_2CT_x) MXenes are employed as the SA material, fabricated with two different host polymers: polyvinyl alcohol (PVA) and polyethylene oxide (PEO). These Nb_2CT_x-based thin-film SAs were integrated into the laser cavity by sandwiching them between fiber ferrules; they successfully generated Q-switched operation in a thulium-doped fiber laser (TDFL) system within the near-MIR region. This work presents the first reported implementation of a Q-switched TDFL using Nb_2CT_x-PEO as a saturable absorber. The demonstrated Q-switched operation achieved pulse widths of 1.995 and 1.942 μs for Nb_2CT_x-PEO and Nb_2CT_x-PVA, respectively, both significantly shorter than the 13.3–4.4 μs range reported in earlier studies, highlighting the effectiveness of these MXene-based SAs for pulsed laser generation at 2 μm.

Keywords: niobium carbide, MXene, mid-infrared laser, Q-switched pulse generation, thulium-doped fiber

1. Introduction

Research on 1.5 μm fiber lasers has primarily focused on the wavelength regions used in fiber-optic communications, including the short-band (1460–1530 nm), conventional-band (1530–1565 nm), and long-band (1565–1625 nm). Recently, attention has shifted toward fiber lasers operating at longer wavelengths, particularly around 2 μm. Although optical fibers exhibit higher transmission loss at this wavelength,

2 μm fiber lasers offer promising opportunities beyond telecommunications. Notably, the 2 μm wavelength aligns with the absorption peak of water molecules, allowing for a penetration depth of approximately 0.6 mm, which is an ideal characteristic for surgical applications using lasers [1]. These lasers enhance safety and are considered "eye-safe" since the radiation is absorbed by the eye before it reaches the retina [2]. Moreover, this wavelength lies within an atmospheric transmission window, making it suitable for applications such as remote sensing and free-space optical communication.

In pulsed 2 μm laser generation, passive approaches commonly make use of saturable absorbers. Numerous nanomaterials have been studied as potential saturable absorbers in pulsed fiber lasers, such as transition metal dichalcogenides (TMDs), topological insulators (TIs), black phosphorus (BP), graphene, semiconductor saturable absorber mirrors (SESAMs), and carbon nanotubes (CNTs). Each of these materials offers distinct advantages and limitations. SESAMs exhibit strong absorption but involve complex fabrication [3]. CNTs feature ultrafast recovery times but suffer from limited absorption due to tube size [4]. Graphene offers a broad operational bandwidth [5] but is hindered by its zero bandgap and low modulation depth. Although BP is readily produced and has a wide nonlinear optical response, it deteriorates upon contact with oxygen or moisture [6]. TIs such as Bi_2Se_3 and Bi_2Te_3 possess broad absorption bandwidths, though they have a low damage threshold [7]. TMDs like MoS_2 and $MoTe_2$ offer unique absorption features, but their fabrication processes are often complex [8].

In this chapter, we introduce a novel nanomaterial, MXene, which stands out as a strong candidate for a saturable absorber owing to its distinctive optical, physical, and thermal properties.

2. MXene-based saturable absorber

In recent years, a new class of two-dimensional (2D) material known as MXene has garnered significant attention because of its exceptional optical and optoelectronic properties, including a tunable bandgap and the ability to respond across a broad range of wavelengths. MXenes have been widely used in various applications within nanoscience and photonics, owing to their exceptional chemical and mechanical stability. It also has high electrical and magnetic conductivity, which can enhance the Li-ion storage capacity. In 2017, Jhon et al. reported the first femtosecond pulsed laser with an MXene-based saturable absorber [9], which opened new avenues for exploring MXenes in ultrafast pulse generation. MXene, known as transition metal carbides, nitrides, and carbonitrides, has a general formula of $M_{n+1}X_nT_x$, where M is a transition metal, X is carbon or nitrogen, and T_x represents surface terminations or functional groups that are bonded to the outer M layers, such as fluorine ions (–F), chlorine ions (–Cl), oxygen ions (–O), and hydroxyl (–OH), which depend on the synthesis method [10].

There are a few methods for the synthesis of MXene material, and the preparation method will significantly impact the performance of the pulsed laser. The fabrication of MXene is separated into two approaches: the top-down method and the bottom-up method. The top-down method mostly refers to obtaining the MXene by etching the raw material, while the bottom-up method utilizes the chemical reaction between the transition metal element and other materials that contain elements of carbon and nitrogen. Selective acid etching of the MAX phase is the most common method in the top-down method. The acid-like HF solution can be used to etch the MAX phase to obtain MXene that is accompanied by surface functional groups like –F, –Cl, –O, and –OH [11]. For the bottom-up approach, the chemical vapor deposition (CVD) method is widely used in the fabrication of 2D materials. For example, Mo_2C MXene is obtained by Xu with the CVD method as they mixed the CH_4 into the molten MoCu alloy and created Mo_2C crystals [12].

After the MXene is obtained, it can be incorporated into the laser cavity to function as a saturable absorber. The sandwich structure and evanescent-wave coupling are the two main approaches to integrating the MXene into the fiber laser cavity. The sandwich structure normally utilizes a polymer film that is placed between two fiber ferrules. The polymer film is created by combining MXene with polyvinyl alcohol (PVA) or polyethylene oxide (PEO). Thus, the light that travels in the fiber laser cavity can interact with the polymer film. For the evanescent-wave coupling, a d-shaped fiber or tapered fiber is required. The MXene solution is then deposited on the d-shaped fiber or tapered fiber. The light traveling in the fiber can leak from the d-shaped fiber or tapered fiber and interact with the MXene. The drawback of the sandwich structure is that the polymer film may not withstand the high power of the intracavity light. This drawback can be covered by the evanescent-wave coupling, but the evanescent-wave coupling cannot guarantee sufficient reaction of the MXene and the leaked light [13].

2.1 Fabrication of the MXene-based saturable absorber

All MXenes used in this experiment were prepared by selectively etching the MAX phase material. The etching solution was first prepared by dissolving 0.8 g of lithium fluoride (LiF) in 10 mL of 9 M hydrochloric acid (HCl), with continuous magnetic stirring until full dissolution was achieved. To prevent vigorous bubbling from the exothermic reaction, 0.5 g of Nb_2AlC MAX phase powder was carefully added over 5 minutes under stirring. The solution was then maintained at room temperature (~23°C) for 24 hours to ensure complete etching. After etching, the clear liquid (supernatant) was removed, leaving Nb_2CT_x residue at the bottom. To neutralize the acidic mixture, it was washed with deionized (DI) water and centrifuged at 3500 rpm for 5 minutes. This washing and centrifugation were repeated several times, adding 150 mL of DI water each time, until the pH reached around 5. The final Nb_2CT_x residue was vacuum-dried at 120°C for 24 hours to obtain Nb_2CT_x MXene powder.

Next, the polyethylene oxide (PEO) and polyvinyl alcohol (PVA) were used to form films with MXene powders. For the PEO film, a PEO solution was prepared by dissolving 1 g of PEO powder (900,000 Mw, Sigma Aldrich) in 120 mL of deionized water under stirring at 200°C. Once fully dissolved, Nb_2CT_x MXene powder was added in a 2.5:2.5 ratio (PEO:MXene). This mixture was ultrasonicated for 1 hour, then transferred to a petri dish and dried in air at 24°C. Finally, a self-standing SA film formed, which could be easily peeled off from the petri dish. Using this method, Nb_2CT_x-PEO films were obtained.

For the PVA film, 1 g of PVA powder (40,000 Mw, Sigma Aldrich) was mixed with 120 mL of DI water and stirred at 200°C until a clear solution formed. The MXene powder was then added in a 2.5:2.5 ratio. After one hour of ultrasonic treatment, the mixture was poured into a petri dish and air-dried at room temperature. After the mixture was dried, it formed a self-standing SA film and could be easily peeled off from the petri dish. Finally, the Nb_2CT_x-PEO and Nb_2CT_x-PVA saturable absorbers were obtained.

2.2 Characterization of MXene-based saturable absorber

The material characteristics of MXene-based saturable absorbers were investigated using field emission scanning electron microscopy (FESEM) and energy-dispersive X-ray spectroscopy (EDX), performed with a Zeiss Auriga Crossbeam FIB with an integrated EDX detector. Additionally, the balanced twin-detector method was used to determine the nonlinear optical absorption of the saturable absorbers. The data obtained from this method were graphically represented using a fitting curve based on the saturation model equation applied to the experimental dataset.

2.2.1 Characterization of the Nb_2CT_x-PEO saturable absorber

Figure 1(a) shows the FESEM image of Nb_2CT_x-PEO, where Nb_2CT_x MXene flakes, approximately 1 μm in size, are observed to be distributed evenly within the PEO matrix. The EDX spectrum of the Nb_2CT_x-PEO is presented in **Figure 1(b)**. Characteristic peaks of carbon (C), oxygen (O), fluoride (F), aluminum (Al), silicon (Si), and niobium (Nb) confirm the element composition of the Nb_2CT_x-PEO sample. Since the Nb_2CT_x MXene is synthesized by selectively etching the MAX phase, some residual impurities remain. The aluminum is detected in the sample because the selective etching process did not completely remove the Nb_2AlC MAX phase. However, the percentage of aluminum is only 1.87%, indicating that aluminum has been successfully removed. The presence of fluoride and oxygen is due to both fluoride and oxygen being functional groups of the Nb_2CT_x MXene. Furthermore, silicon originated from the Si-based substrate.

There are some essential parameters for a saturable absorber, such as saturable absorption, non-saturable absorption, saturable intensity, absorption wavelength, and recovery time. Additionally, the balanced twin-detector technique was employed to characterize the nonlinear optical absorption behavior. The equation of the absorption coefficient is shown below [14]:

$$\alpha(I) = \frac{\alpha_s}{1 + \frac{I}{I_s}} + \alpha_{ns} \tag{1}$$

Where α_s, α_{ns}, I, and I_s represent the saturable absorption, non-saturable absorption, intensity of incident light, and saturation intensity, respectively. **Figure 1(c)** illustrates the nonlinear optical absorption response of Nb_2CT_x-PEO, which exhibits standard nonlinear saturable absorption behavior, with a modulation depth of 19.67%, a non-saturable loss of 74.7%, and a saturation intensity of 0.451 kW/cm^2. The relatively low saturable intensity and high non-saturable loss are influenced by the pulsed laser's characteristics in the balanced twin-detector method. Additionally, these values may also be influenced by the fabrication quality of the Nb_2CT_x-PEO saturable absorber. The elevated non-saturable loss introduces greater cavity loss, which favors Q-switched operation over mode-locked operation.

2.2.2 Characterization of the Nb_2CTx-PVA saturable absorber

Similarly, for the Nb_2CT_x-PVA saturable absorber, the FESEM image at 200 magnification and a 20 kV energy threshold is shown in **Figure 2(a)**. It reveals a few

DOI: http://dx.doi.org/10.5772/intechopen.1013103

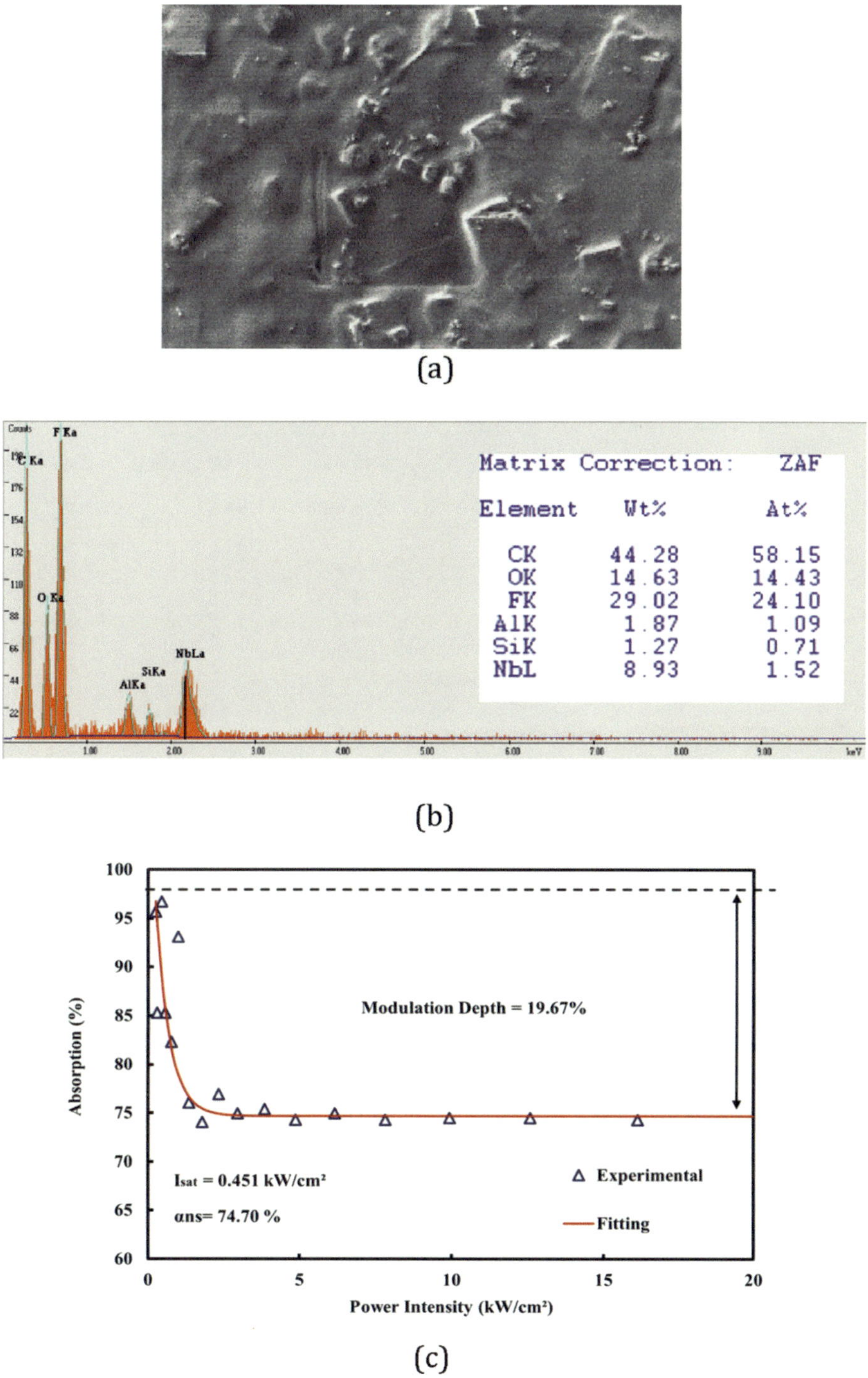

Figure 1.
Characteristics of the Nb_2CT_x-PEO film in the FESEM image (a), EDX profile (b), and curve fitting of nonlinear optical absorption (c).

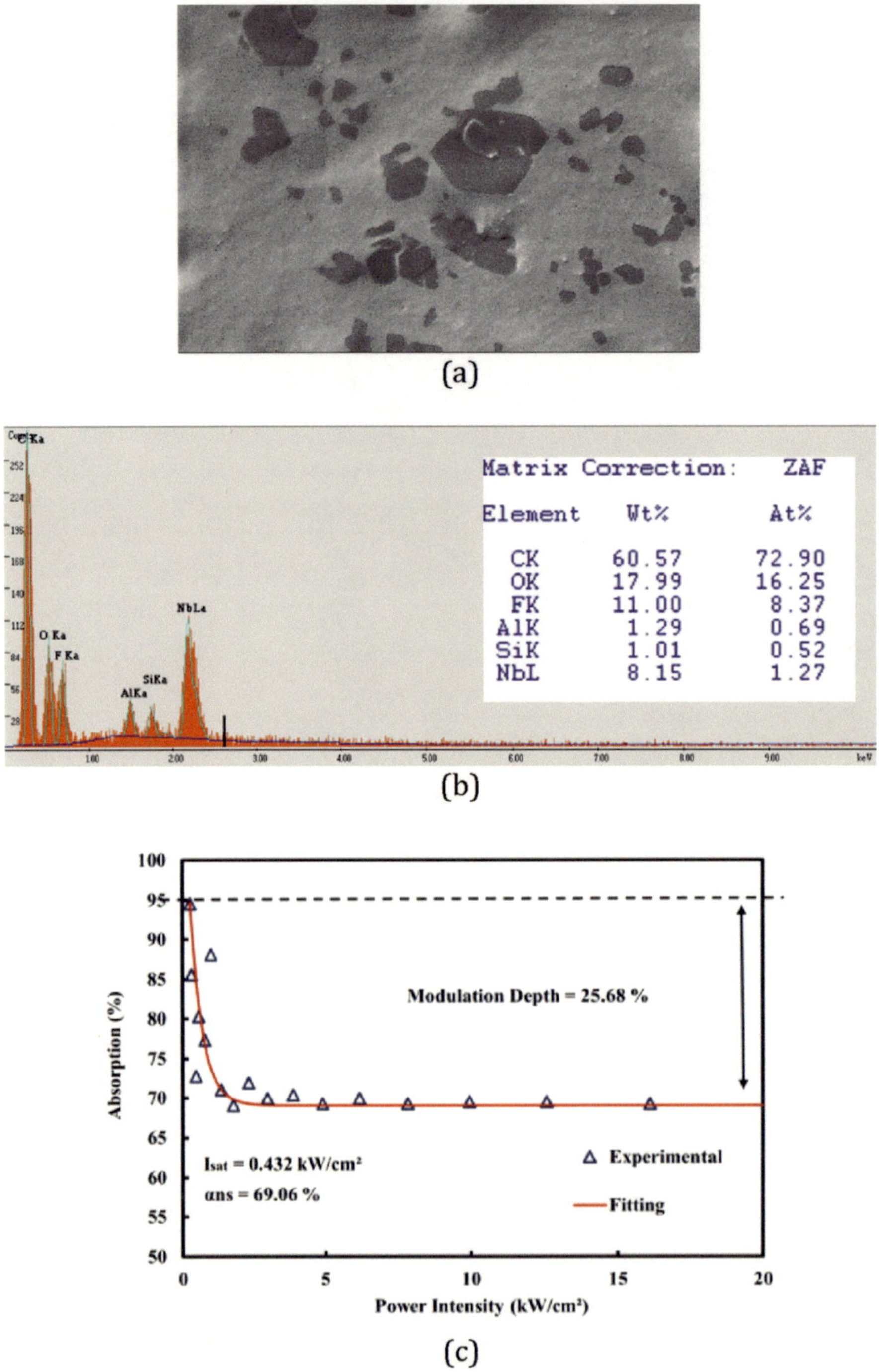

Figure 2.
Fabricated Nb_2CT_x-PVA film characterization. FESEM image (a), EDX profile (b), and nonlinear optical absorption curve (c).

crystals, proving the presence of Nb_2CT_x-PVA. EDX analysis shown in **Figure 2(b)** detected elements such as C, O, F, Al, Si, and Nb. Oxygen and fluoride originate from surface functional groups, while a small amount of aluminum (1.29%) indicates the successful removal of aluminum by the selective etching method. The Nb_2CT_x-PVA had a modulation depth of 25.68%, a non-saturable loss of 69.06%, and a saturation intensity of 0.432 kW/cm^2 shown in **Figure 2(c)**.

3. Thulium-doped fiber as the laser gain medium

A novel thulium-doped fiber (TDF) is used as the laser gain medium to generate laser emission in the near-MIR region. The TDF was fabricated at the Central Glass and Ceramic Research Institute, India. It has a composition of SiO_2-GeO_2-Al_2O_3-HfO_2-Y_2O_3-Bi_2O_3 and was created from a preform prepared using the modified chemical vapor deposition (MCVD) process. The fabricated preform then underwent the fiber drawing process. Next, the CVD process enables the doping of P_2O_5 into silica, along with the integration of glass modifiers such as Al_2O_3, Y_2O_3, HfO_2, and Bi_2O_3 using the solution doping technique. In this process, a germano-silica layer deposited inside a silica tube at a temperature of 13,500°C was soaked for 60 minutes in an alcohol-water mixture (1:5 ratio) containing $Al(NO_3)_3$: $9H_2O$, YCl_3: $6H_2O$, $HfOCl_2$: $8H_2O$, and $Bi(NO_3)_3$: xH_2O. After the solution was drained, an inert nitrogen gas flow at ambient temperature was used to dry the soaked layer. The soaked layer was then heated to approximately 9000°C in an oxygen atmosphere to produce the corresponding oxides from the halide salts. Through sintering of the porous layer at 1400–1850°C, these oxides, such as GeO_2, Al_2O_3, Y_2O_3, HfO_2, and Bi_2O_3, were fused into the silica glass as a transparent layer and subsequently collapsed at a high temperature above 2000°C to form the preform. A fiber was subsequently drawn using a fiber drawing tower, with an on-line resin coating applied at 2000°C. HfO_2 was selected because of its high refractive index and transparency across a broad wavelength range (0.3–10 μm). The SiO_2-GeO_2-Al_2O_3-HfO_2-Y_2O_3-Bi_2O_3 glass was chosen because of its promising potential for high-power laser applications requiring a high optical damage threshold. Within the silica glass matrix, hafnium ions interact with more than four oxygen atoms, resulting in the creation of non-bridging oxygen (NBO) atoms in the network. This structure facilitates the integration of active elements like rare earths and bismuth without inducing clustering effects. The doping of HfO2 is also able to vaguely modify the structure of silica core glass, assisting the dispersion of active bismuth co-dopants.

Designed for single-mode operation, the TDF has a cladding-to-core diameter of 125 ± 0.5/6.5 μm and a numerical aperture (NA) of approximately 0.20. The NA of the fiber was measured as 0.20 using a fiber analyzer. **Figure 3(a)** presents a microscopic view of the fiber cross-section. Using electron probe microanalyses (EPMA), the doping levels of various elements along the fiber diameter were measured. **Figure 3(b)** shows the dopant distribution curve within the fiber core. The composition of the fiber core glass is 6.15 wt% of Al_2O_3, 2.2 wt% of Y_2O_3, 0.47 wt% of Tm_2O_3, 2.12 wt% of GeO_2, 1.57 wt% of HfO_2, and 0.05 wt% of Bi_2O_3. Absorption spectra of the TDF were obtained in the visible and near-infrared (NIR) regions, as illustrated in **Figure 3(c)** and **(d)**. In the visible region, peak absorptions were observed at wavelengths of

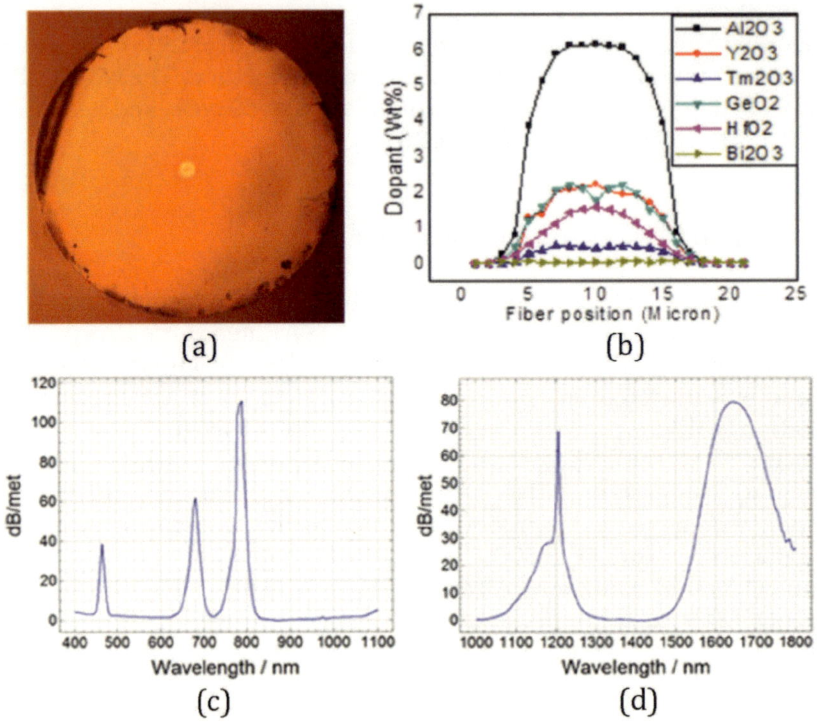

Figure 3.
(a) TDF cross-sectional microscopic view, (b) the concentration profiles of different dopants along the fiber core diameter, (c) the absorption profile in the visible region, and (d) the NIR region.

795, 1200, and 1550 nm with corresponding absorption coefficients of 110, 68, and 30 dB/m. As depicted in **Figure 3(d)**, the NIR region exhibited a peak absorption at 1640 nm with a value of 80 dB/m.

4. Construction of the laser cavity

The laser cavity configuration of the Q-switched thulium-doped fiber laser (TDFL) incorporating an MXene-based saturable absorber is shown in **Figure 4**. The pump source used was a homemade erbium-ytterbium co-doped fiber laser (EYDFL) with an operating wavelength of 1552 nm. The power of the homemade laser was varied using a laser diode controller (Arroyo Instruments: 4320), with a maximum pump power of 997.26 mW. The 1550 nm input port of a 1550/2000 wavelength division multiplexer (WDM) was connected to the 1552 nm pump source. A 5 m-long TDF, serving as the gain medium for the 2 μm laser cavity, was connected to the 2000 nm output port of the WDM. The other end of the TDF was connected to an intermediate port, which in turn was connected to the input of a 90:10 output coupler. The saturable absorber was inserted at the intermediate port. To complete the TDFL ring cavity, the 90% port of the output coupler was connected to the 2000 nm input port of the WDM. The 10% port was used to extract a portion of the laser output for measurement purposes.

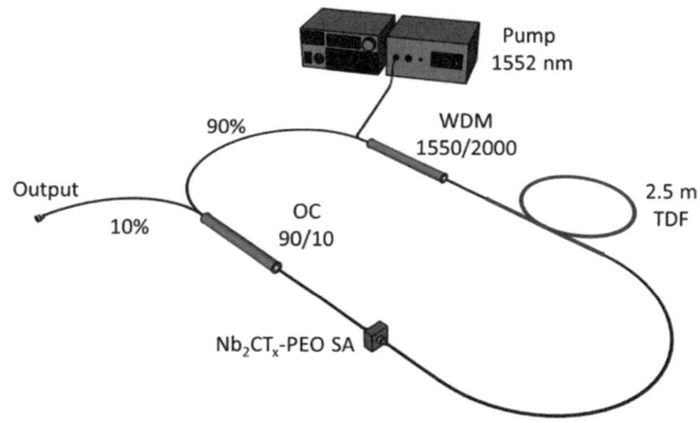

Figure 4.
Schematic of the experimental setup for the TDFL cavity.

A spectrometer (Miriad S3) was used for analyzing the output spectrum. Additionally, an oscilloscope (GWINSTEK: MDO-2302AG) was employed for time-domain analysis. The oscilloscope also supports frequency domain analysis, as it includes a built-in 1 GHz radio frequency (RF) spectrum analyzer. A 12.5 GHz InGaAs photodetector (EOT: ET-5000F) was connected to the oscilloscope to convert optical signals into electrical signals for measurement. Finally, an optical power meter (Thorlabs PM100D) was used to measure the output power. The laser cavity did not require a polarization controller, as Q-switched pulses were generated immediately upon the insertion of the saturable absorber.

5. Performance of the Q-switched pulse laser using the niobium carbide MXene saturable absorber

The Q-switching performance of the TDFL using two different niobium carbide MXene saturable absorbers was investigated. First, in the absence of any saturable absorber, continuous-wave (CW) laser operation was initially established at a pump power of 415.57 mW. Thereafter, the Nb_2CT_x-PEO and Nb_2CT_x-PEO saturable absorbers were each inserted between two fiber ferrules in the TDFL cavity. Stable Q-switching was achieved at pump powers of 626.13 and 611.09 mW for the Nb_2CT_x-PEO and Nb_2CT_x-PVA saturable absorbers, respectively, and remained stable as the power increased further to 731.41 and 776.53 mW. An increase in pump power led to reductions in both the pulse interval and pulse width, as presented in **Figure 5(a)** and **(b)**, consistent with typical Q-switched behavior. The removal of the saturable absorber caused the TDFL to operate in CW mode with chaotic, noise-like pulses, most likely arising from self-pulsing effects inside the cavity [15]. As shown in **Figure 6(a)**, the output spectrum of the TDFL incorporating the Nb_2CT_x-PEO SA exhibits a center wavelength of 1888.36 nm at a pump power of 731.41 mW. In comparison, **Figure 6(b)** displays a center wavelength of 1964.121 nm at 776.53 mW for the TDFL with the Nb_2CT_x-PVA SA. In the absence of a saturable absorber, the CW laser produced a center wavelength of 1966.113 nm, as illustrated in the insets of **Figures 6(a)** and **6(b)**. With the inclusion of the SA, the blue shift in the center wavelength can be explained by the laser's tendency to shift into a higher gain region to compensate for the additional loss. The output spectrum of the Q-switched laser with Nb_2CT_x-PEO SA was lower than that of the Nb_2CT_x-PVA SA, likely due to higher

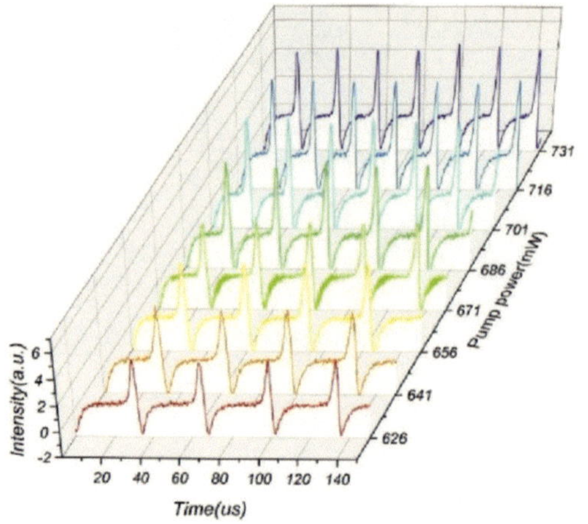

(a)

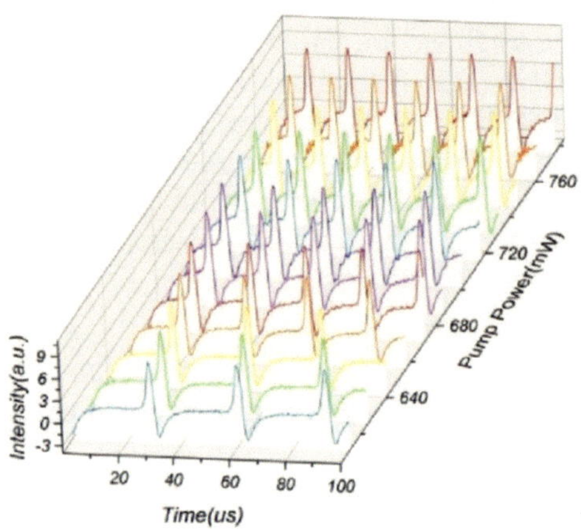

(b)

Figure 5.
Q-switched pulse trains recorded at varying pump powers for (a) Nb_2CT_x-PEO and (b) Nb_2CT_x-PVA saturable absorbers.

insertion losses associated with PEO-based SA. Notably, the Q-switched pulse train was no longer observed once the pump power reached 806.61 mW for both SAs, suggesting an over-saturation effect. Nevertheless, when the pump power was lowered to 731.41 mW for the Nb_2CT_x-PEO SA and 776.53 mW for the Nb_2CT_x-PVA SA, the Q-switched pulses returned, suggesting that both SAs remained undamaged at the maximum pump power.

The pulse trains at various pump powers are shown in **Figure 5(a)** and **(b)** for the Nb_2CT_x-PEO and Nb_2CT_x-PVA saturable absorbers, respectively. In both cases, the pulse trains exhibit a bright-dark pulse pair, with a weaker dark pulse trailing each bright pulse. This phenomenon arises from the nonlinear interaction between the two orthogonal polarization components of the pulses. The polarization-independent nature of both the TDF and the saturable absorbers enables the coexistence of these orthogonal polarization components within the laser cavity, facilitating coherent coupling and the formation of dark pulses [16]. Within the TDFL cavity, loss

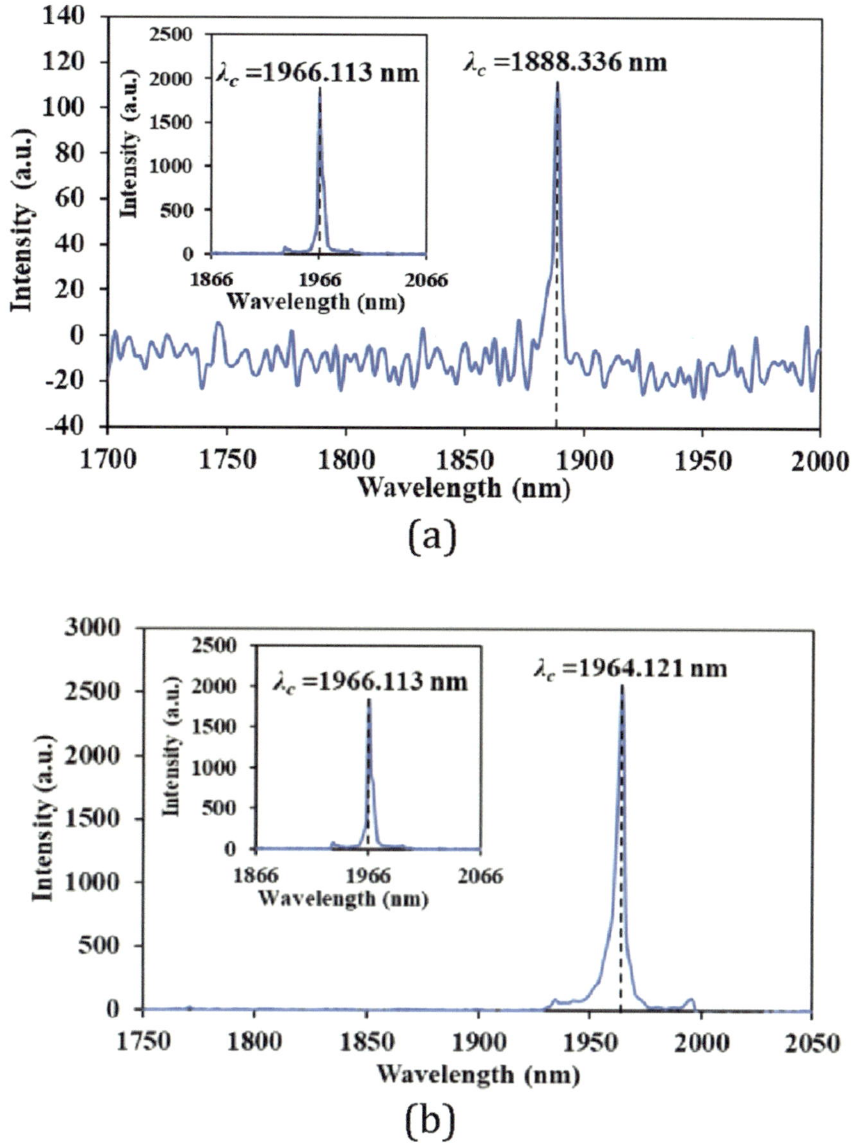

Figure 6.
Output spectra of the Q-switched TDFL incorporating (a) Nb_2CT_x-PEO and (b) Nb_2CT_x-PVA saturable absorbers. The insets display the output spectrum of the CW laser operation in the absence of saturable absorbers.

modulation and the nonlinear optical response of the Nb_2CT_x-PEO and Nb_2CT_x-PVA films facilitate the emergence of bright-dark pulse dynamics [17]. This interaction significantly contributes to the observed laser performance.

Figure 7(a) and **(b)** present the radio frequency (RF) spectra of the Q-switched laser at pump powers of 686.3 and 716.37 mW for Nb_2CT_x -PEO SA and Nb_2CT_x-PVA SA, respectively. For the Nb_2CT_x-PEO SA, the RF spectrum shows a fundamental frequency at 39 kHz, with the inset indicating a signal-to-noise ratio (SNR) of approximately 45 dB, confirming stable Q-switched operation. In contrast, the Nb_2CT_x-PVA SA demonstrates a fundamental frequency of 57 kHz with a higher SNR

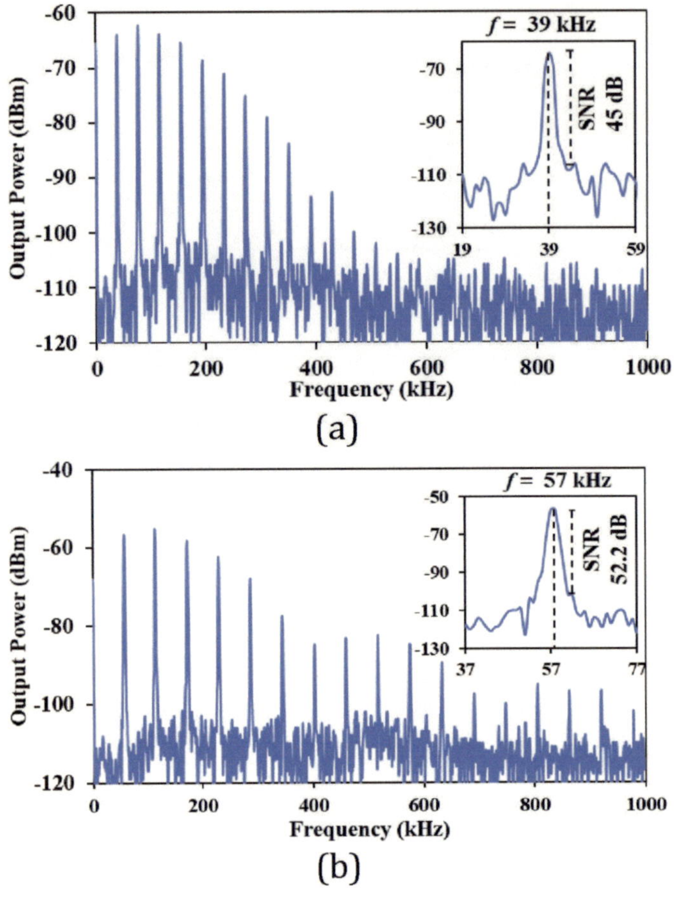

Figure 7.
RF spectra with insets displaying the signal-to-noise ratio (SNR) and fundamental frequency for (a) Nb_2CT_x-PEO SA at pump power of 686.26 mW and (b) Nb_2CT_x-PVA SA at 716.37 mW.

of 52.2 dB, also indicating stable performance. Both configurations exhibit multiple harmonic components extending up to 1000 kHz. Notably, mode-locked pulses were not observed in this cavity setup, likely due to high intracavity loss. As such, to enable mode-locking operation, further optimization of the laser cavity is required.

To further investigate the Q-switched laser performance, over pump powers of 626.13–731.41 mW for the Nb_2CT_x-PEO SA and 611.09–716.37 mW for the Nb_2CTx-PVA SA, the repetition rate and pulse width were measured systematically. As shown in **Figure 8(a)**, the Q-switched TDFL with Nb_2CT_x-PEO SA exhibited an initial repetition rate of 29.53 kHz at the threshold pump power, which gradually increased to 46.7 kHz as the pump power was raised to 731.41 mW. Simultaneously, the pulse width decreased from 2.950 to 1.995 μs. **Figure 8(b)** illustrates the corresponding behavior for the Nb_2CT_x-PVA SA, where the reduction in pulse width from 2.415 to 1.956 μs corresponded with an increase in repetition rate from 31.64 to 68.41 kHz. This inverse relationship between pump power and pulse width, along with the increase in repetition rate, is the characteristic of typical Q-switched laser dynamics. At higher pump power, greater energy is delivered to the TDF, enabling faster saturation of the SA, which in turn leads to higher pulse repetition rates and narrower pulse widths.

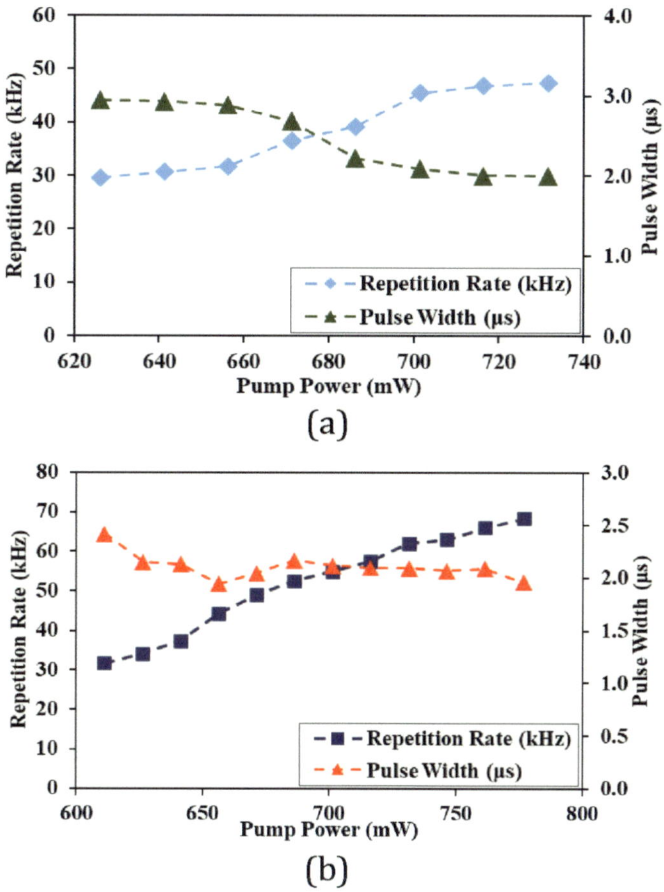

Figure 8.
Variation of repetition rate and pulse width with pump power for (a) Nb_2CT_x -PEO and (b) Nb_2CT_x -PVA saturable absorbers.

Figure 9(a) presents the variation of peak power with respect to pump power for the Nb_2CT_x-PEO saturable absorber. The peak power shows a slight rise from 3.650 to 5.286 mW as the pump power increases from 626.13 to 731.41 mW. In contrast, for the Nb_2CT_x-PVA SA, shown in **Figure 9(b)**, the peak power initially increases from 50.517 to 54.732 mW as the pump power rises from 611.09 to 626.13 mW. However, a subsequent decline in peak power is observed from 50.517 to 41.477 mW as the pump power continues to increase to 716.37 mW. This decline is attributed to the presence of dark pulses, which absorb energy that would normally contribute to the peak power.

Figure 9(c) shows how the output power and pulse energy vary with pump power for the Nb_2CT_x-PEO SA. As the pump power is increased from 626.13 to 731.41 mW, the average output power rises from 0.318 to 0.500 mW. Meanwhile, the pulse energy drops to 10.546 nJ following a slight increase from 10.769 to 11.102 nJ as the pump power increases. For the Nb_2CT_x-PVA SA, depicted in **Figure 9** (d), the average output power increases from 3.86 to 5.55 mW, whereas the pulse energy decreases from 121.997 to 81.128 nJ as the pump power rises from 611.09 to 776.53 mW. The increase in average output power is caused by the gain supplied by the TDF in the laser cavity. However, the reduction in pulse energy for both SAs is likely due to increased losses introduced by components within the TDFL cavity.

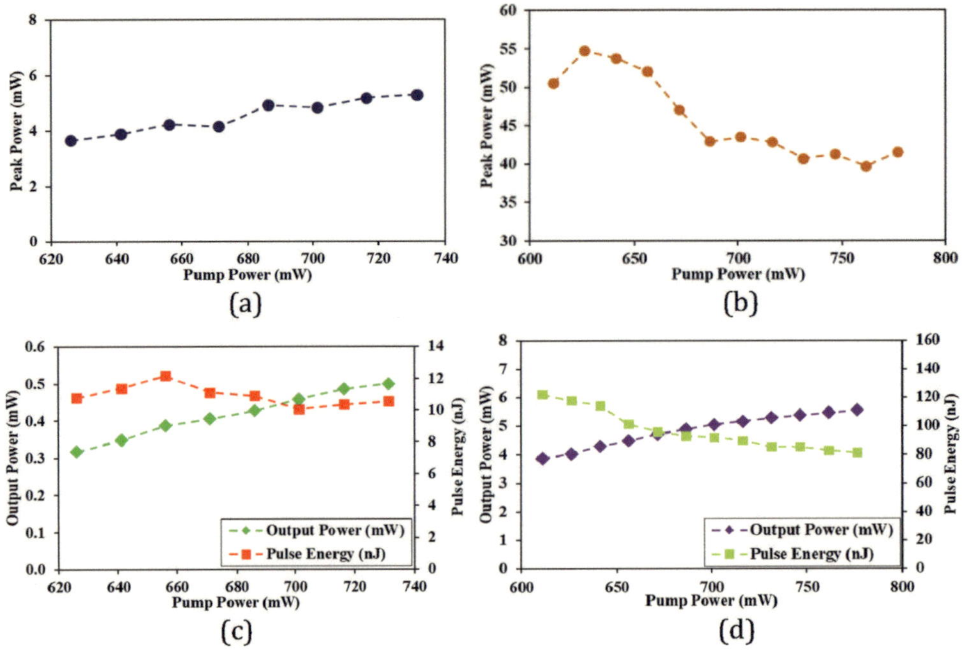

Figure 9.
Variation of peak power with pump power for (a) Nb_2CT_x-PEO SA and (b) Nb_2CT_x-PVA SA. Output power and pulse energy as functions of pump power for (c) Nb_2CT_x-PEO SA and (d) Nb_2CT_x-PVA SA.

A comparative overview of prior research on pulsed fiber lasers with Nb_2C-based saturable absorbers is provided in **Table 1**. Nb_2C demonstrates its versatility by generating pulses at multiple wavelengths, including 1.3, 1.5, and 2 μm. This study investigated Q-switched operation using a thin film of Nb_2CT_x-PEO and Nb_2CT_x-PVA integrated into a thulium-doped fiber laser cavity. As far as we are aware, this represents the first demonstration of a Q-switched fiber laser utilizing Nb_2CT_x-PEO as SA in this configuration. Moreover, this study demonstrates Q-switched operation with

Saturable absorber	Gain medium	Wavelength (nm)	Laser operation	Repetition rate	Pulse width	Method	SNR (dB)	Ref.
Nb_2C	EDF	1559.98	Mode-locked	12.54 MHz	603 fs	Tapered fiber	>45	[18]
Nb_2C	TDF	1882.13	Mode-locked	6.28 MHz	2.27 ps	Tapered fiber	>60	[18]
Nb_2C-PVA	EDF	1558.73	Q-switched	75.19 kHz	13.3 μs	Thin film	53.8	[19]
Nb_2C	EDF	1559	Mode-locked	14.12 MHz	0.77 ps	Tapered fiber	66.5	[20]
Nb_2C	TDF	1944	Mode-locked	9.35 MHz	1.67 ps	Tapered fiber	>52	[21]
Nb_2C	THDF	1950	Mode-locked	11.76 MHz	1.34 ps	Tapered fiber	>64	[21]
Nb_2C-PVA	BDF	1314	Q-switched	13.8 kHz	7.58 μs	Thin film	>60	[22]
Nb_2C-PVA	HDF	2079.5	Q-switched	20.5 kHz	4.4 μs	Thin film	57	[23]

Saturable absorber	Gain medium	Wavelength (nm)	Laser operation	Repetition rate	Pulse width	Method	SNR (dB)	Ref.
Nb_2CT_x-PEO	TDF	1888.3	Q-switched	46.7 kHz	1.995 μs	Thin film	45	This work
Nb_2CT_x-PVA	TDF	1964.12	Q-switched	68.41 kHz	1.942 μs	Thin film	52.2	This work

Table 1.
Comparative overview of pulsed laser performance employing niobium carbide saturable absorbers.

pulse widths of 1.995 and 1.942 μs for Nb_2CT_x -PEO SA and Nb_2CT_x -PVA SA, respectively, significantly narrower than those reported in previous works, which ranged from 13.3 to 4.4 μs. These findings highlight the potential of thin-film-based Nb_2CT_x -PEO and Nb_2CT_x -PVA saturable absorbers for generating mode-locked pulses with TDF in the 2.0 μm region.

6. Conclusion

In this chapter, the niobium carbide-based MXene was successfully synthesized *via* a selective etching method. For the Nb_2CT_x MXene, two different host polymers, polyethylene oxide (PEO) and polyvinyl alcohol (PVA), were used to form composite films, namely Nb_2CT_x-PEO and Nb_2CT_x-PVA. The fabricated MXenes were characterized using field emission scanning electron microscopy (FESEM), energy-dispersive X-ray spectroscopy (EDX), and modulation depth measurements. Both saturable absorbers exhibited modulation depth values suitable for use in Q-switched laser applications. The thin-film SAs were then incorporated into a thulium-doped fiber laser (TDFL) cavity through a connecting port. Q-switching performance of the TDFL using both Nb_2CT_x-PEO and Nb_2CT_x-PVA SAs was successfully demonstrated. Each SA produced stable Q-switching pulse trains that followed the typical characteristics of Q-switched operation. However, it was observed that the Nb_2CT_x-PEO SA resulted in higher insertion loss compared to Nb_2CT_x-PVA SA, as indicated by the lower output intensity of the Q-switched spectrum. Overall, the performance of Nb_2CT_x-PVA SA is better than Nb_2CT_x-PEO SA in terms of repetition rate, pulse width, and insertion loss. In the case of bright-dark pulse trains, the peak power can be improved, as the power contributed to the dark pulse is significant, and it limits the peak power growth of the bright pulse. This constraint may be addressed in the future by optimizing the polarization state using a polarization controller, which could enable the generation of bright Q-switched pulses with higher pulse energy and peak power. Furthermore, upgrading components within the TDFL cavity is essential for enabling additional laser regimes, such as mode-locking, dark-pulse generation, and supercontinuum generation, which would broaden the range of potential applications. This is attributed to the high intracavity losses, which have inhibited laser generation in other operation regimes. Besides that, implementing a dual wavelength pumping scheme at 795 and 1550 nm could compensate for these losses and support more diverse laser behaviors [24]. Notably, this work represents the first reported implementation of a Q-switched TDFL using Nb_2CT_x-PEO as a saturable absorber. The demonstrated Q-switched operation achieved pulse widths of 1.995 and 1.942 μs for Nb_2CT_x-PEO and Nb_2CT_x-PVA, respectively, both significantly shorter than the 13.3–4.4 μs range reported in previous works.

Acknowledgements

This work was supported by the University of Wollongong Malaysia under the Internal Research Grant 2024 with grant number (UOWMRG/2024/004).

Author details

Cheng Xiau San*, Lip Zhan Hong and Tan Sin Jin
School of Engineering, University of Wollongong Malaysia, Shah Alam, Selangor, Malaysia

*Address all correspondence to: xs.cheng@uow.edu.my

References

[1] Boone C. Medical applications are a surgical fit for 2 μm lasers. Laser Focus World. 2022;**59**(01):50-51. Available from: https://www.laserfocusworld.com/lasers-sources/article/14286709/medical-applications-are-a-surgical-fit-for-2-m-lasers. [Accessed: December 14, 2022]

[2] Scholle K, Lamrini S, Koopmann P, Fuhrberg. 2 μm laser sources and their possible applications. In: Pal B, editor. Frontiers in Guided Wave Optics and Optoelectronics. London, UK, London: InTech; 2010. DOI: 10.5772/39538

[3] Huang C, Xiao H, Xu J, Tang Y. Selectively mode-locked high-order-mode 2-μm thulium fiber laser. Optics Communications. 2019;**440**:95-99. DOI: 10.1016/j.optcom.2019.01.070

[4] Liu HH, Chow KK, Yamashita S, Set SY. Carbon-nanotube-based passively Q-switched fiber laser for high energy pulse generation. Optics & Laser Technology. 2013;**45**:713-716. DOI: 10.1016/j.optlastec.2012.05.005

[5] Peng X, Yan Y. Graphene saturable absorbers applications in fiber lasers. Journal of the European Optical Society-Rapid Publications. 2021;**17**(1):16. DOI: 10.1186/s41476-021-00163-w

[6] Wang T, Jin X, Yang J, Wu J, Yu Q, Pan Z, et al. Oxidation-resistant black phosphorus enable highly ambient-stable ultrafast pulse generation at a 2 μm Tm/Ho-doped fiber laser. ACS Applied Materials & Interfaces. 2019;**11**(40):36854-36862. DOI: 10.1021/acsami.9b12415

[7] Li P, Zhang G, Zhang H, Zhao C, Chi J, Zhao Z, et al. Q-switched mode-locked Nd:YVO_4 laser by topological insulator Bi_2Te_3 saturable absorber. IEEE Photonics Technology Letters. 2014;**26**(19):1912-1915. DOI: 10.1109/LPT.2014.2341832

[8] Ahmad H, Suthaskumar M, Tiu ZC, Zarei A, Harun SW. Q-switched erbium-doped fiber laser using $MoSe_2$ as saturable absorber. Optics & Laser Technology. 2016;**79**:20-23. DOI: 10.1016/j.optlastec.2015.11.007

[9] Jhon YI, Koo J, Anasori B, Seo M, Lee JH, Gogotsi Y, et al. Metallic MXene saturable absorber for femtosecond mode-locked lasers. Advanced Materials. 2017;**29**(40):1702496. DOI: 10.1002/adma.201702496

[10] Naguib M, Kurtoglu M, Presser V, Lu J, Niu J, Heon M, et al. Two-dimensional nanocrystals produced by exfoliation of Ti_3 AlC_2. Advanced Materials. 2010;**23**(37):4248-4253. DOI: 10.1002/adma.201102306

[11] Zhou J, Zha X, Chen FY, Ye Q, Eklund P, Du S, et al. A two-dimensional zirconium carbide by selective etching of Al_3C_3 from nanolaminated $Zr_3Al_3C_5$. Angewandte Chemie International Edition. 2016;**55**(16):5008-5013. DOI: 10.1002/anie.201510432

[12] Xu C, Wang L, Liu Z, Chen L, Guo J, Kang N, et al. Large-area high-quality 2D ultrathin Mo_2C superconducting crystals. Nature Materials. 2015;**14**(11):1135-1141. DOI: 10.1038/nmat4374

[13] Cheng Y, Lyu W, Wang Z, Ouyang H, Zhang A, Sun J, et al. MXenes: Synthesis, incorporation, and applications in ultrafast lasers. Nanotechnology. 2021;**32**(39):392003. DOI: 10.1088/1361-6528/ac0d7e

[14] Ahmad H, Albaqawi HS, Yusoff N, Bayang L, Kadir MZBA, Yi CW.

Tunable passively Q-switched erbium-doped fiber laser based on $Ti_3C_2T_x$ MXene as saturable absorber. Optical Fiber Technology. 2020;**58**:102287. DOI: 10.1016/j.yofte.2020.102287

[15] Wang M, Huang S, Zeng Y-J, Yang J, Pei J, Ruan S. Passively Q-switched thulium-doped fiber laser based on oxygen vacancy MoO_{3-x} saturable absorber. Optical Materials Express. 2019;**9**(11):4429. DOI: 10.1364/OME.9.004429

[16] Zhang W, Zhan L, Xian T, Gao L. Generation of bright/dark pulses in an erbium-doped fiber laser mode-locked with glycerin. Journal of Lightwave Technology. 2019;**37**(15):3756-3760. DOI: 10.1109/JLT.2019.2919808

[17] Omar S, Musa B, Rosol AHA, Jusoh Z, Yasin M, Harun SW. Generation of bright-dark Q-switched pulse pair in thulium-doped fiber laser cavity with titanium aluminum carbide absorber. Microwave and Optical Technology Letters. 2023;**65**(12):3333-3340. mop.33837. DOI: 10.1002/mop.33837

[18] Gao L, Ma C, Wei S, Kuklin AV, Zhang H, Ågren H. Applications of few-layer Nb_2C MXene: Narrow-band photodetectors and femtosecond mode-locked fiber lasers. ACS Nano. 2021;**15**(1):954-965. DOI: 10.1021/acsnano.0c07608

[19] Ahmad H, Yusoff N, Albaqawi HS, Reduan SA, Thambiratnam K. Niobium carbide (Nb_2C) MXene as a saturable absorber to assist in the generation of a wavelength tunable passively Q-switched fiber laser. Laser Physics Letters. 2021;**18**(6):065101. DOI: 10.1088/1612-202X/abf83c

[20] Ahmad H, Ramli R, Yusoff N, Reduan SA, Zamzuri AK, Thambiratnam K. Performance of Nb_2C MXene coated on tapered fiber as saturable absorber for the generation of mode-locked erbium-doped fiber laser. Infrared Physics & Technology. 2021;**114**:103647. DOI: 10.1016/j.infrared.2021.103647

[21] Ahmad H, Ramli R, Ismail NN, Aidit SN, Yusoff N, Samion MZ. Passively mode locked thulium and thulium/holmium doped fiber lasers using MXene Nb_2C coated microfiber. Scientific Reports. 2021;**11**(1):11652. DOI: 10.1038/s41598-021-90978-x

[22] Ahmad H et al. 1.3 μm passively Q-switched bismuth doped fiber laser using Nb_2C saturable absorber. Optical Materials. 2021;**116**:111087. DOI: 10.1016/j.optmat.2021.111087

[23] Ahmad H, Ismail NN, Aidit SN, Reduan SA, Samion MZ, Yusoff N. 2.08 μm Q-switched holmium fiber laser using niobium carbide-polyvinyl alcohol (Nb_2C-PVA) as a saturable absorber. Optics Communication. 2021;**490**:126888. DOI: 10.1016/j.optcom.2021.126888

[24] Lai WJ, Tan YC. Dual-wavelength pumping (793 nm + 1600 nm) for thulium doped silica fiber laser. In: 2018 23rd Opto-Electronics and Communications Conference (OECC). New York, NY, United States: Institute of Electrical and Electronics Engineers (IEEE); 2018. pp. 1-2. DOI: 10.1109/OECC.2018.8729848

Chapter 5

Pulse Generation through Multimode Interference

Sulaiman Wadi Harun, Yu Chen, Ahmad Haziq A. Rosol, Zian Cheak Tiu and Retna Apsari

Abstract

Pulsed lasers, particularly ultrashort pulsed lasers, have extensive applications in fields such as spectroscopy, telecommunications, atmospheric measurements, surgery, bio-imaging, and material processing. This chapter explores the use of the single-mode-multimode-single-mode (SMS) structure as an effective means for developing simple, compact, and cost-efficient Q-switched and mode-locked fiber lasers. The SMS configuration functions as a clever artificial saturable absorber (SA), enabling the generation of ultrashort pulsed lasers via the Kerr effect within nonlinear multimode interference. By fusion-splicing single-mode fibers to both ends of a specific length of multimode fiber, the SMS structure is formed. Integrating this SMS structure into a ring erbium-doped fiber laser (EDFL) and fine-tuning parameters using a polarization controller led to the successful generation of ultrashort pulsed lasers. A series of experiments were conducted in both anomalous and normal dispersion regimes to evaluate the potential of various SMS structures for Q-switched and mode-locked pulse generation. In the net anomalous dispersion regime, Q-switched fiber lasers were successfully demonstrated using step-index multimode fibers (SIMF) within the SMS structure. Additionally, mode-locked EDFLs were achieved with graded-index multimode fibers (GIMF) in the SMS configuration, operating in both conventional soliton and bound soliton regimes. The artificial SA showed a significantly higher damage threshold compared to conventional thin-film absorbers.

Keywords: pulsed fiber lasers, Q-switching, mode-locking, artificial saturable absorber, Kerr effect, nonlinear optics

1. Introduction

Fiber lasers are highly promising across a wide range of scientific and industrial fields, making up a significant portion of the commercial laser market. They are generally classified into two types based on their mode of operation: continuous and pulsed. Pulsed fiber lasers, known for their high pulse energy and short pulse duration, have greatly expanded the application of lasers in areas such as fiber-optic communication, medical devices, laser cutting, and automotive manufacturing.

The generation of pulsed lasers primarily relies on two techniques: Q-switching and mode-locking, both of which can be implemented either actively or passively. Active techniques involve external components, such as acousto-optic or electro-optic modulators, to control the laser cavity. In contrast, passive techniques utilize the inherent properties of the laser cavity to generate short pulses without external components, typically employing saturable absorbers (SAs) to modulate the intra-cavity light, leading to Q-switching or mode-locking [1]. This passive approach is favored for its simplicity, cost-effectiveness, and reliability [2].

SAs play a critical role in determining the performance of Q-switched and mode-locked pulse generation. Natural SAs are typically based on the nonlinear absorption properties of materials, where photons are absorbed within the material. Common natural SAs include semiconductor saturable absorber mirrors (SESAMs) [3], carbon nanotubes (CNTs) [4], transition metal chalcogenides (TMDs) [5], graphene [6], and black phosphorus (BP) [7]. While these materials offer benefits, they also face drawbacks such as high cost, complex fabrication processes, narrow operational bandwidths, and low damage thresholds, motivating researchers to explore the development of artificial SAs (ASAs).

The introduction of ASAs has significantly advanced the development of Q-switched and mode-locked fiber lasers, overcoming the limitations of natural SAs [8]. Thanks to their all-fiber composition, ASAs integrate seamlessly into optical setups and offer a high damage threshold. Unlike their conventional counterparts, which have limited operational bandwidths, ASAs exhibit versatility across a wide range of wavelengths. They also display favorable nonlinear optical characteristics, such as high modulation depth, low non-saturable loss, and low saturation intensities, which improve laser stability, efficiency, and pulse quality.

ASAs rely on nonlinear optical effects, rather than direct photon absorption, to achieve pulse generation. Various ASA types include nonlinear polarization rotation (NPR) [9], nonlinear polarization evolution (NPE) [10], nonlinear optical loop mirror (NOLM) [11], nonlinear amplifying loop mirror (NALM) [12], and single-mode fiber (SMF)-multimode fiber (MMF)-SMF structures (referred to as SMS structures) [8]. NPR and NPE use polarization phenomena and polarization instabilities within the laser cavity to achieve mode-locking, while NOLM and NALM utilize fiber loops to generate the same effect. SMS structures exploit the Kerr effect through nonlinear multimode interference (NL-MMI).

Among these, the SMS structure is the most recent and least explored type of ASA. However, there remains a lack of systematic studies on the SMS structure's potential for generating various Q-switched and mode-locked pulses. In this chapter, we aim to experimentally explore the capabilities of SMS structures, utilizing step-index multimode fiber (SIMF) and graded-index multimode fiber (GIMF) as ASAs for Q-switched and mode-locked pulse generation, respectively.

2. Multimode interference

When light enters an MMF, it excites a unique set of MMF modes, each with its own propagation constant. As these modes propagate, they interact, causing multimode interference (MMI). The resulting field distribution within the MMF at any point is the combination of these modes, showing the complex interactions between them. Self-imaging in MMF, first observed by Allison et al., occurs when the optical field periodically replicates itself [13]. While the superposition of modes generally leads to a complex field, self-imaging can occur at specific points where the excited modes are in phase, meaning the phase difference between them is a multiple of 2π. This condition is described by Zhu [14]. However, finding exact self-imaging positions is difficult due to the different propagation constants of the modes, so quasi-reproduction, where the field is nearly replicated, is often targeted. Larger core MMFs can excite more modes and have longer self-imaging periods, but careful consideration of MMF length is needed to achieve high-quality self-imaging.

At the splice point illustrated in **Figure 1**, N modes are triggered. These excited modes can be mathematically described by the following equation:

$$E_{sm}\left(r,\varnothing,z=0\right)=\sum_{1}^{N}C_n e_n\left(r,\varnothing,z=0\right), \tag{1}$$

Here, $E_{sm}\left(r,\varnothing,z=0\right)$ represents the fundamental mode of the SMF, and $e_n\left(r,\varnothing,z=0\right)$ is the n-th guided mode of the MMF. N refers to the excited modes inside the MMF, and C_n is the mode expansion coefficient. Assuming no mode conversion, the excited modes propagate inside the MMF, and the field $E_{MM}\left(r,\varnothing,z\right)$ along the MMF is given by:

$$E_{MM}\left(r,\varnothing,z\right)=\sum_{1}^{N}C_n e_n\left(r,\varnothing,0\right)e^{-i\beta_n z}=e^{-i\beta_1 z}\sum_{1}^{N}C_n e_n\left(r,\varnothing,0\right)e^{-i(\beta_n-\beta_1)z}, \tag{2}$$

where β_1 and β_n are the propagation constants of the fundamental mode and the n-th excited mode of the MMF, respectively. From Eq. (2), it is clear that self-imaging of the input field occurs when the following condition is met for all N modes:

$$\left(\beta_n-\beta_1\right)z_{self-imaging}=\Delta\beta_n z_{self-imaging}=m_n 2\pi, \tag{3}$$

where m_n is an integer. Eq. (3) indicates that self-imaging will happen when the phase difference between any two excited modes is an integer multiple of 2π. However, since the propagation constants of the excited modes are not integer multiples of each other, achieving perfect self-imaging at a specific position within a practical MMF length (several meters) is nearly impossible. As a result, we typically aim for quasi-reproduction positions where the phase difference between the modes is slightly off from 2π but still results in high-quality self-imaging, such as over 99% reproduction of the input field.

It is important to note that, due to mode orthogonality and on-axis excitation, only $LP_{0,n}$ modes, under the linear polarization approximation, are strongly excited in the MMF. Generally, the larger the core of the MMF, the more modes are excited and the longer the self-imaging period. Therefore, to achieve high-quality self-imaging with large-core MMFs, careful consideration of the MMF length is necessary.

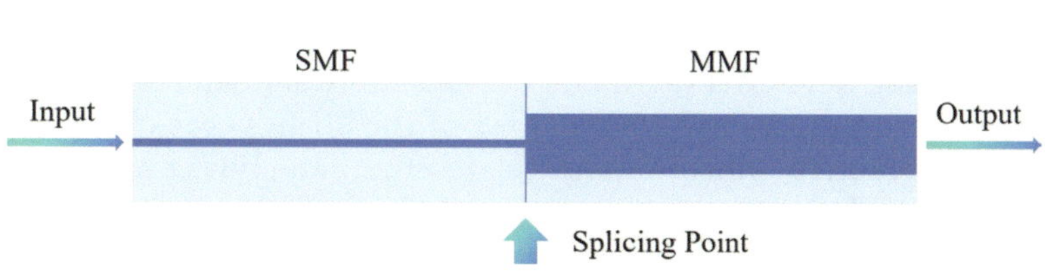

Figure 1.
Schematic drawing of light coupling from an SMF to an MMF.

Consider an MMF of a specific length L, Eq. (3) can be expressed as:

$$\Delta n_{eff,n} L = m_n \lambda_0, \tag{4}$$

where $\Delta n_{eff,n} = \frac{(\beta_n - \beta_1)\lambda_0}{2\pi} = n_{eff,n} - n_{eff,1}$ and λ_0 is the operating wavelength of a specific intensity laser. Eq. (4) clearly shows that self-imaging at the output occurs only at a specific wavelength. This phenomenon is known as the spectral filtering effect of nonlinear multimode interference (NI-MMI). Due to the Kerr effect, β_n and β_1 are intensity-dependent propagation parameters. Therefore, Eq. (4) can be rewritten as:

$$\Delta n_{eff,n}(I) L = m_n \lambda_0, \tag{5}$$

As intensity increases, the operational wavelength can be determined using the following relation:

$$\lambda_{high} = \frac{\Delta n_{eff,n}(I_{high})}{\Delta n_{eff,n}(I_0)} \lambda_0, \tag{6}$$

This equation emphasizes the spectral filtering effect inherent in an SMS structure. For GIMF, Nazemosadat and Mafi [15] proposed using the SMF-GIMF-SMF structure as a spectral modulator or SA. Their numerical analysis showed that due to NL-MMI, the SMF-GIMF-SMF structure exhibits spectral filtering capabilities, selectively attenuating low-power signals while allowing higher-power signals to propagate effectively.

3. Q-switched pulse generation using SIMF-based ASA

In this section, we introduce a novel Q-switched erbium-doped fiber laser (EDFL) that utilizes a SIMF based ASA. SIMF is a type of MMF where the refractive index is uniformly distributed across the core. As light propagates through SIMF, it follows a zigzag path, leading to significant modal dispersion. The large core diameter of SIMF allows multiple modes to propagate simultaneously, causing interference between them and resulting in intensity-dependent transmission similar to a natural SA. Extensive theoretical studies have demonstrated that a SIMF spliced with an SMF structure can function as an ASA for Q-switching operations [14]. However, there are limited experimental demonstrations of Q-switching with SIMF [16]. When laser light transitions from an SMF (with a core diameter of 8.3 μm) to an SIMF (with a core diameter of 105 μm), the large mismatch in core diameters at the splice points can cause mode excitation and coupling. Due to the Kerr effect, which becomes more significant as the input light intensity increases, the refractive index of the SIMF's core changes. At low intensities, the refractive index remains relatively stable, but at higher intensities, the nonlinear refractive index variation alters the behavior of the SIMF's core. This change shifts the conditions under which interference occurs, reducing the coupling of light into certain modes. As a result, the transmission of light at higher intensities increases, making the SMS structure function like an SA.

3.1 Experimental arrangement

The experimental arrangement consists of two main stages. First, we fabricate and characterize the SMS structure. Once the SMS structure is fabricated and characterized, it is integrated into a ring cavity for Q-switching operation. The SMS structure is formed using a 158 mm long SIMF. A schematic representation of the SMS structure is shown in **Figure 2(a)**. The SIMF (Thorlabs, FG105LVA) is fusion spliced with two segments of SMFs (Corning, SMF-28) at each end, using a Fujikura 90S+ splicer. The splice losses were measured to be 0.01 and 0.02 dB, respectively. The SIMF has core and cladding diameters of 105 and 125 μm, while the SMF has core and cladding diameters of 8.2 and 125 μm, respectively. These three fiber segments together form the SMS structure.

To evaluate the SIMF-based ASA's characteristics, both linear and nonlinear characterizations were conducted. A homemade amplified spontaneous emission (ASE) laser was used for linear characterization. **Figure 2(b)** shows the spectral transmission characteristics of the SMS structure, clearly displaying a periodic interference

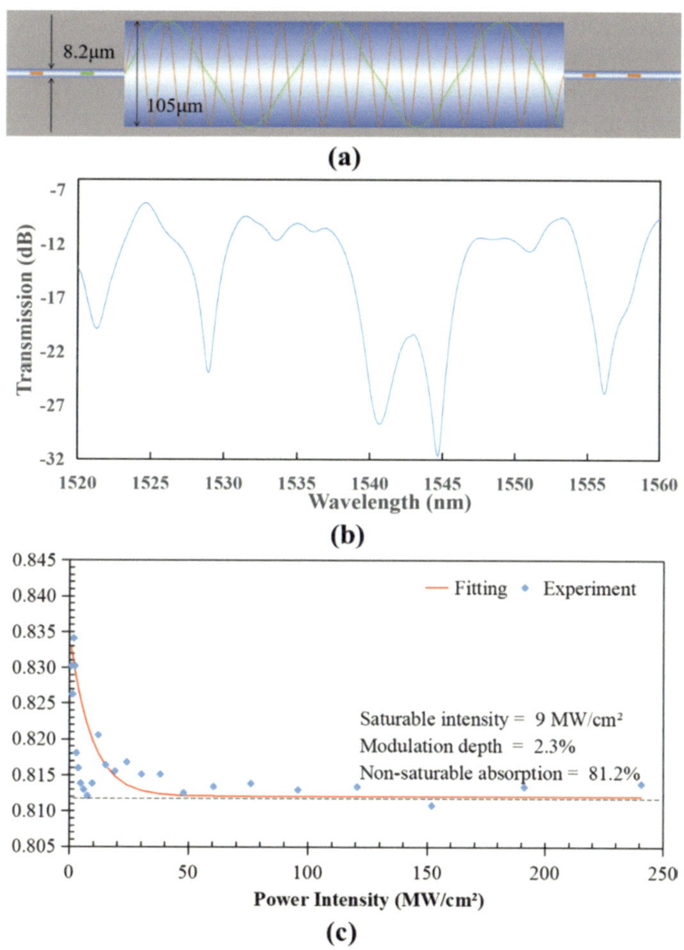

Figure 2.
(a) Schematic drawing, (b) spectral transmission characteristic, and (c) absorption curve of the SIMF-based SMS structure. [Reprinted/Adapted] with permission from [17].

pattern caused by multimode interference from the various excited modes as the laser light propagates from the SMF into the SIMF. A balanced twin-detector system was used to measure the modulation depth and saturation intensity of the SMS structure. The laser source used for this was a homemade laser operating at a center wavelength of 1567 nm, with a pulse width of 700 fs and a repetition rate of 21.73 MHz. The nonlinear transmission curve was fitted using the following equation:

$$\alpha(I)=\frac{\alpha_S}{1+\frac{I}{I_S}}+\alpha_{NS}, \tag{7}$$

where $\alpha(I)$, α_S, I, I_S and α_{NS} denotes absorption rate, modulation depth, input light intensity, saturation intensity, and non-saturable absorption. The experiment result is shown in **Figure 2(c)**. It shows that modulation depth is 2.3%, which shows a not significant change before and after the SMS structure. This might be because of the poor coupling quality in between the SIMF and SMFs. Another possible reason might be because the competing nonlinear effects like SPM or XPM interfere with the saturable absorption process, thus reducing the modulation depth. The saturable intensity is 9 MW/cm^2, showing a low saturation threshold. The non-saturable absorption is 81.2%, which is very high non-saturable absorption. This is likely due to the bending and splicing loss.

After characterizing the SMS structure, it was fusion-spliced into an EDFL cavity (**Figure 3**) to function as an ASA. The continuous wave laser diode (LD) (II-VI, Model LC96Z400-74) provides a maximum output power of 343 mW at 975 nm. The LD is spliced to a 980/1550 nm wavelength division multiplexer (WDM). A 2-meter-long erbium-doped fiber (EDF, Fibercore I25), serving as the gain medium, is connected to the cavity following the WDM. The output from the EDF is directed to a polarization-insensitive isolator to ensure unidirectional laser operation. The SMS structure is placed after the isolator. The SIMF section of the ASA is positioned in an in-line

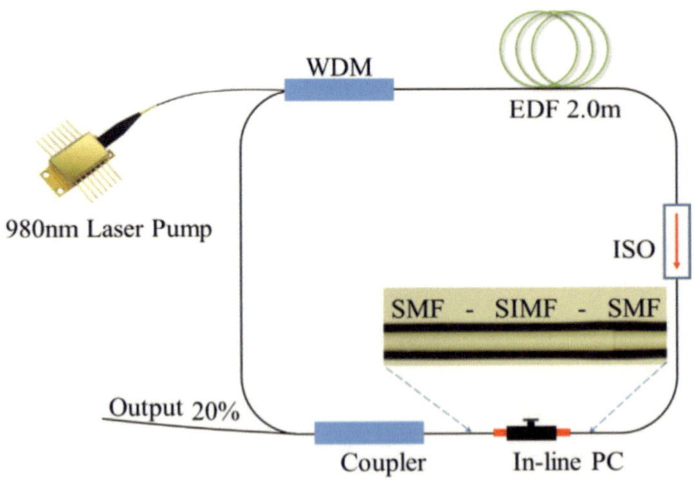

Figure 3.
Schematic diagram of the experimental setup for Q-switched pulse generation in the ring cavity, utilizing SIMF as the ASA. The inset displays a microscope image of the SMS structure and its splicing point. [Reprinted/Adapted] with permission from [17].

polarization controller (PC) to minimize SIMF bending. The in-line PC can also apply the required stress and strain on the SIMF to control the polarization of the oscillating laser when necessary. An 80:20 coupler is positioned after the SMS structure to provide feedback to the cavity and output port. 20% of the oscillating light is tapped from the ring cavity for further analysis. The optical spectrum is analyzed using an optical spectrum analyzer (OSA, Anritsu, MS9710C) with a resolution of 0.02 nm. A high-speed photodetector (ET-3500F) connected to a digital oscilloscope (GW Instek, GDS-3352) monitors the real-time signal in the temporal domain. The stability of the Q-switching operation is assessed in the frequency domain using a radiofrequency (RF) spectrum analyzer (Anritsu, MS2830A). The total length of the cavity is approximately 9.6 meters, which includes the gain medium, SIMF, and SMFs.

3.2 Q-switching performance

Q-switched laser pulses were observed with a threshold input pump power of 115 mW, which was relatively high due to the significant loss in the SMS structure. The Q-switched operation continued up to the maximum pump power of 343 mW. Notably, no pulses were generated when the SMS structure was removed from the cavity. The spectral characteristics of the Q-switched pulse laser were carefully analyzed, as shown in **Figure 4(a)**. Operating at 1530 nm, the laser maintained a stable wavelength across different input pump powers. This wavelength was specifically

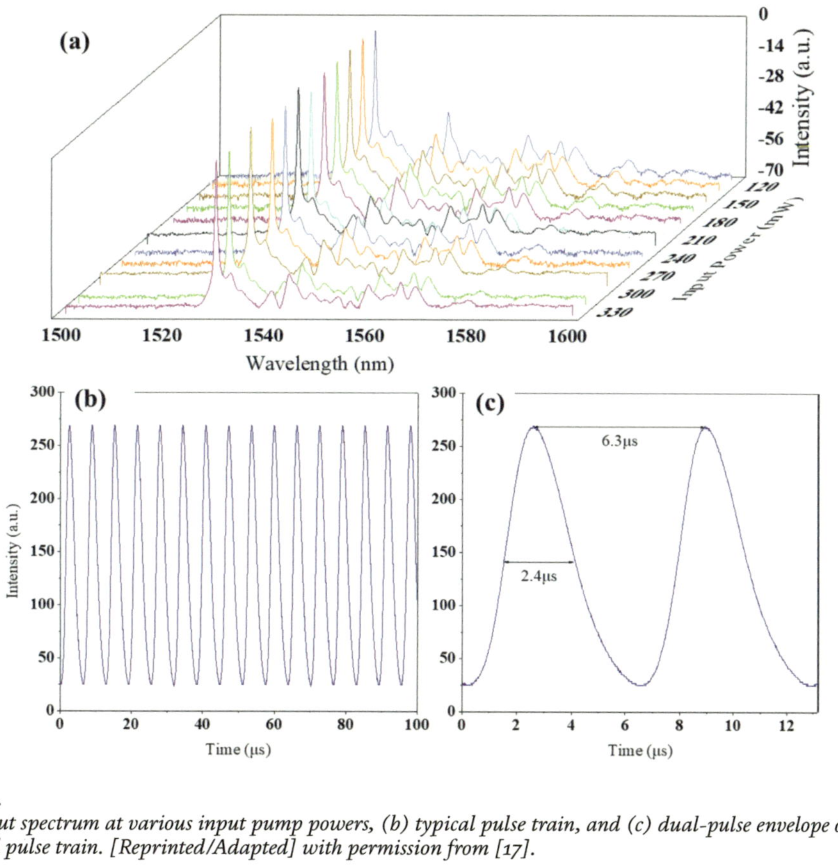

Figure 4.
(a) Output spectrum at various input pump powers, (b) typical pulse train, and (c) dual-pulse envelope of the expanded pulse train. [Reprinted/Adapted] with permission from [17].

chosen for its optimal gain, which compensates for the losses within the SMS structure. However, the self-imaging effect induced by the SIMF's large core diameter led to less-than-ideal output quality, with the spectrum displaying multiple lower intensity peaks ranging from 1530 to 1580 nm. A typical pulse waveform at 343 mW is shown in **Figure 4(b)**, which demonstrates a stable pulse train with no significant intensity fluctuations. **Figure 4(c)** presents the dual-pulse envelope of the enlarged pulse train, showing a pulse period of 6.3 μs and a duration of 2.5 μs, corresponding to a repetition rate of 157.7 kHz.

The RF spectrum of the output laser at 343 mW pump power is shown in **Figure 5(a)**. It reveals a signal-to-noise ratio (SNR) of 72 dB at the fundamental frequency of 157.7 kHz, indicating the high stability of the laser. At least 13 harmonics were observed within a frequency range of 1.8 MHz, further confirming the stability of the Q-switching operation. **Figure 5(b)** illustrates the variation in repetition rate and pulse width with respect to different pump powers. As the pump power increased from 115 to 343 mW, the repetition rate rose from 87.4 to 157.7 kHz, while the pulse width decreased from 4.2 to 2.5 μs. This behavior is typical of a Q-switched fiber laser, where the pulse width reduction results from gain compression in the laser cavity at higher pump powers. **Figure 5(c)** shows the relationship between the average output power and pulse energy of the Q-switched laser as a function of pump power. The output power increased linearly from 0.61 to 2.35 mW as the pump power rose from

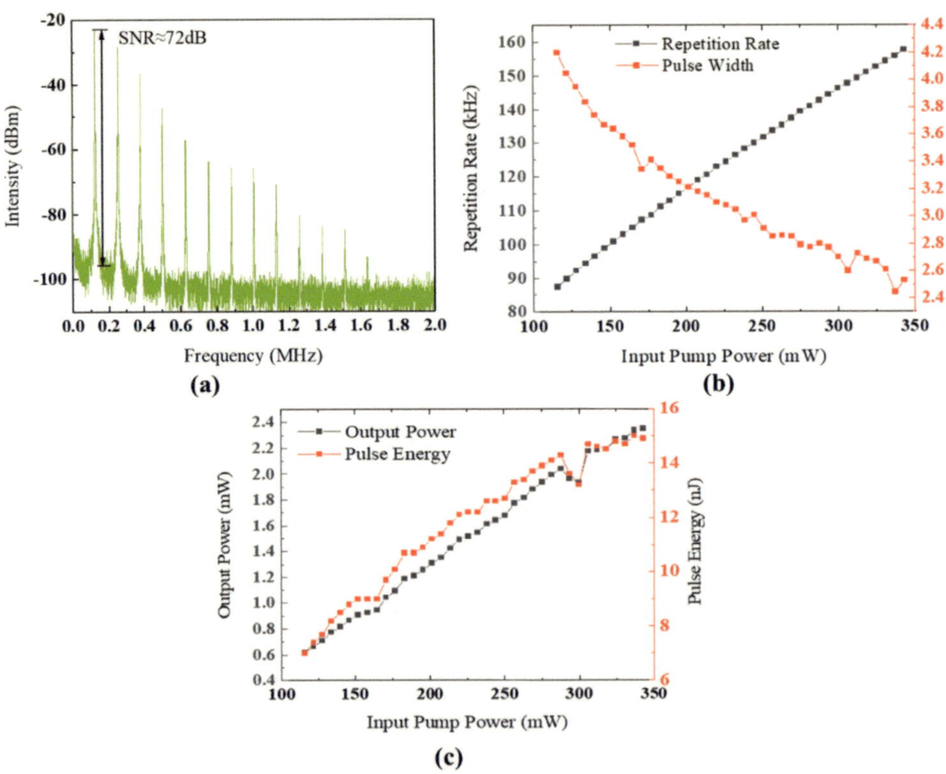

Figure 5.
(a) RF spectrum, (b) repetition rate and pulse width with respect to the pump power, and (c) average output power and pulse energy with respect to the pump power for the SMS based Q-switched EDFL. [Reprinted/Adapted] with permission from [17].

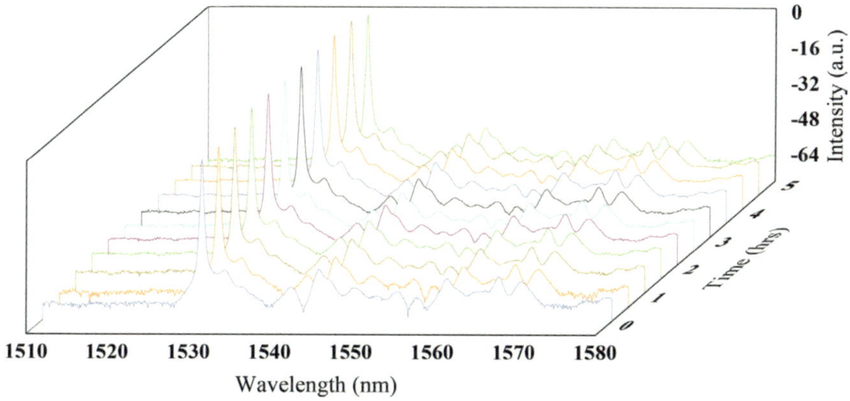

Figure 6.
Long-term stability of the Q-switching operation with SMS structure. [Reprinted/Adapted] with permission from [17].

115 to 343 mW, while the pulse energy grew from 7.0 to 15.0 nJ in the same range. The slope efficiency is approximately 0.8%, which is relatively low due to the high cavity losses. To assess the long-term stability of the Q-switched laser, the cavity was operated at the maximum pump power of 343 mW for 5 hours in a laboratory environment, with spectrum data recorded every half hour. The results, shown in **Figure 6**, demonstrate that the spectra remained largely unchanged throughout the measurement period, indicating excellent stability. After operating for over 24 hours in the laboratory, no noticeable degradation in performance was observed, further confirming the long-term stability of the Q-switching operation using the proposed SMS structure as an ASA.

In summary, we have successfully demonstrated a stable Q-switched laser pulse by integrating an SMF-SIMF-SMF structure as an ASA in the EDFL cavity. This configuration is capable of withstanding high pump power and generating single pulse energies up to 15 nJ, which is an order of magnitude higher than those typically achieved by mode-locked fiber lasers, highlighting the high pulse energy potential of Q-switched pulses. Due to the pump power limitations of our current setup, we were unable to determine the maximum power threshold of this structure. However, given its all-fiber design, it is expected to possess a high damage threshold, potentially supporting nanosecond Q-switched pulses at even higher pump powers. Additionally, the system exhibits excellent long-term stability under stable conditions, indicating the potential for practical use of this structure as an ASA in real-world applications. Furthermore, this proposed SMF-SIMF-SMF structure is simple, cost-effective, and provides reliable laser stability, making it an attractive option for Q-switched pulse generation. However, it is important to note that while the Q-switched operation demonstrates good long-term stability, the SMS structures were found to be more sensitive to environmental fluctuations compared to thin-film SAs, which could lead to long-term stability issues in practical applications.

4. Mode-locking operation using GIMF-based ASA

Optical solitons represent a distinct phenomenon in mode-locked fiber lasers, arising when the interplay between gain, loss, dispersion, and nonlinearity achieves a delicate balance. However, this equilibrium can be disrupted by changes in the polarization controller or increases in pump power, leading to the splitting of a single pulse into multiple pulses with lower peak energy. This effect, known as the peak-power limiting effect, results in pulses that interact with one another through both repulsive and attractive forces [18, 19]. When these forces cause the pulses to bind together, maintaining fixed time separations and phase differences, they form bound solitons or higher-order solitons [20, 21]. Bound solitons can be further classified into tightly bound solitons, or soliton molecules, and loosely bound solitons, depending on the strength of their interactions. Tightly bound solitons exhibit stable, tightly fixed time separations, and corresponding spectrum modulation periods, while loosely bound solitons display more variable time separations and unstable modulation periods [19]. Key features of bound solitons include the modulation period in the optical spectrum and the appearance of multiple symmetrical peaks in the autocorrelation trace. The intensity of these peaks is proportional to the intensity ratio of the solitons in the bound configuration. Researchers have established a crucial relationship between the spectrum modulation period and the time separation of the peaks in the autocorrelation trace, as given by the equation:

$$\Delta T = \frac{\lambda_0^2}{c \cdot \Delta\lambda}, \tag{8}$$

where ΔT is the time separation, $\Delta\lambda$ is the spectrum modulation period, c is the speed of light, and λ_0 is the center wavelength.

The study of bound solitons has advanced both theoretically and experimentally. Malomed et al. [22] and Akhmediev [23] predicted the formation of bound-state pulses with stable soliton spacing at phase differences of 0, π, and π/2 using coupled nonlinear Schrödinger equations (NLSE) and the quintic complex Ginzburg–Landau equation (GLE). Tang et al. [24] explained the formation of multiple solitons as a result of the peak-power limiting effect by analyzing light propagation in weakly birefringent fibers using coupled NLSE. Sakaguchi et al. [25] reported the transformation of different types of bound solitons through simulations of a near-zero-dispersion laser cavity using the complex GLE with cubic-quintic nonlinearity and a third-order dispersion term. More recently, Kulagin et al. [26] used NLSE with a non-Hermitian term to simulate nonlinear optical systems, accounting for environmental effects. Although these theoretical studies and simulations have provided insights into potential soliton behaviors, they are limited by the scope of data and may not fully replicate experimental observations.

In experimental work, Tang et al. [27] observed the generation of bound soliton pulses in fiber lasers using NPE technology. Seong et al. [28] detected the formation of fourth-order bound solitons in a figure-of-eight cavity. Additionally, Pang et al. [29] and He et al. [30] demonstrated all-optical bit storage in fiber lasers using optically mechanically bound solitons. Zhu et al. [19] achieved controllable tightly and loosely bound solitons by employing a hybrid structure with no-core fiber (NCF) and graded-index multimode fiber (GIMF). Wang et al. [31] utilized a similar NCF-GIMF structure for saturable absorption and generated dissipative solitons and bound solitons with adjustable pulse widths and tunable spectral bandwidths under net-normal dispersion conditions. Ren et al. [32] experimentally observed conventional and bound solitons in a 2.0 μm Tm-doped fiber laser using a GIMF-based ASA. These experimental investigations illustrate the potential for generating various types of bound solitons using ASAs.

GIMF, a type of multimode fiber with a refractive index that decreases gradually from the core axis to the core-cladding interface, significantly reduces modal dispersion compared to SIMF. Due to multimode interference, GIMF-based SMS structures support intensity-dependent transmission similar to natural SAs. Theoretical and experimental studies of GIMF-based SMS structures as artificial SAs for mode-locking are well-established [33–36], demonstrating their effectiveness as candidates for mode-locked lasers. In this section, we experimentally demonstrate mode-locking operation with both conventional and bound solitons using a simple GIMF-based SMS structure as an artificial SA.

Soliton pulse generation occurs due to the Kerr effect, which induces NL-MMI in the SMS structure. Notably, our ASA successfully generates both conventional solitons and bound-state soliton pulses. This study represents the first simultaneous observation of tightly and loosely bound solitons within an EDF optical path, utilizing a single segment of GIMF as the SA component. Moreover, this SA structure is simple, easy to fabricate, and exhibits exceptionally low self-starting mode-locking pump powers of 29 and 54 mW, respectively, for conventional and bound-state solitons. Compared to traditional material-based SAs, ASAs offer the advantages of wavelength independence and longer lifetimes, making them highly advantageous for soliton pulse generation.

4.1 Preparation and characterization of GIMF-based ASA

The SMS structure was initially fabricated and characterized. A segment of GIMF (YOFC, OM4) was directly spliced to standard SMFs (Corning, SMF-28) on both ends to create the SMF-GIMF-SMF structure. The GIMF had a length of 45 mm, with core and cladding diameters of 50 and 125 μm, respectively. **Figure 7(a)** shows a schematic diagram of the SMF-GIMF-SMF structure along with microscope images of both splicing points.

When light transitions from an SMF to a GIMF, it experiences multimode interference (MMI). This occurs because of the significant difference in core diameters at the splice point, which leads to the excitation and coupling of multiple light modes [37]. As a result, a periodic interference pattern appears in the spectral transmission characteristics, as shown in **Figure 7(b)**. This pattern is a consequence of the MMI induced by the various excited modes during the laser light's propagation through GIMF. The fiber's varying refractive index profile excites multiple modes, causing interference effects that manifest as the observed periodic pattern in the transmission spectrum. While this phenomenon can be utilized in applications like multimode interferometers for sensing or filtering, it can also function as an ASA. To evaluate the modulation depth and saturation intensity of the SMS-based artificial SA, a balanced twin-detector system was employed [36]. A custom mode-locked fiber laser was used as the light source. The nonlinear transmission curve was fitted using the following equation:

$$T(I) = 1 - q \cdot e^{-I/I_{sat}} - a_{ns}, \tag{9}$$

Here, q is the modulation depth, I_{sat} is the saturation intensity, and a_{ns} represents the non-saturable loss. The experimental results are shown in **Figure 1(c)**, revealing q = 4.6%, I_{sat} = 19.33 MW/cm^2, and a_{ns} = 27.4%. These findings indicate that the absorption properties of the SMS structure vary with changes in incident light intensity. Compared to the SIMF-based SMS structure, this configuration demonstrates a higher modulation depth, making it more suitable for mode-locking applications.

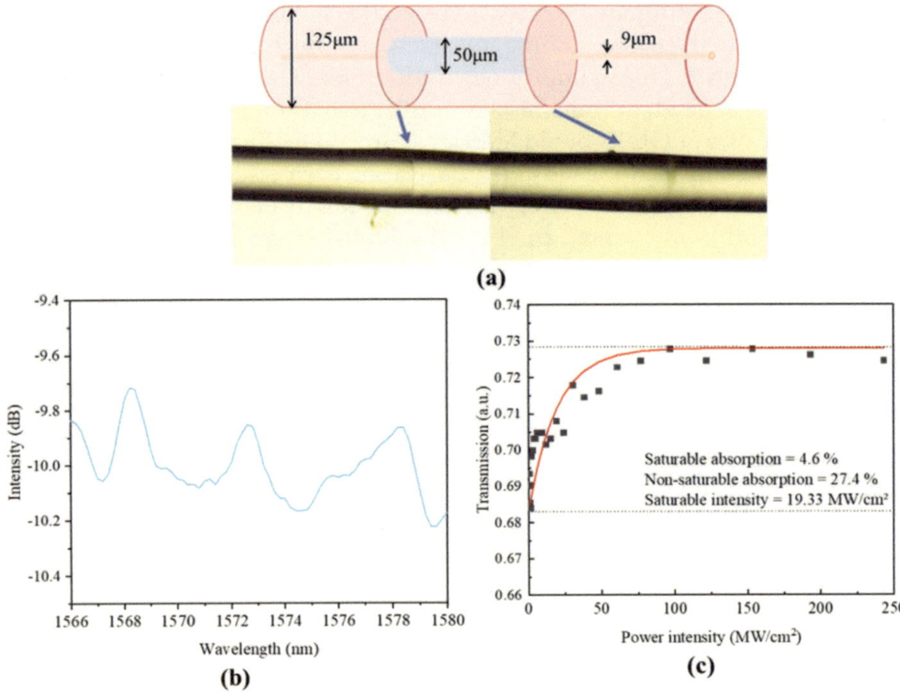

Figure 7.
(a) The schematic diagram of the SMS structure with the image of splicing point at both ends of GIMF, (b) The spectral transmission characteristic of the SMS structure, and (c) Nonlinear transmission curve of the SMS structure-based SA.

4.2 Configuration of the mode-locked EDFL

Figure 8 presents the schematic diagram of the proposed mode-locked fiber laser, which features a ring cavity configuration with the GIMF-based SMS structure functioning as an ASA. The setup is similar to the previous Q-switched laser, with the exception of the SA component. To prevent bending of the GIMF, which could alter the self-imaging length, the SMS structure is enclosed within an in-line PC. The laser's wavelength and time-domain performance were characterized using an OSA and a high-speed photodetector coupled to a digital oscilloscope, respectively. An optical power meter with an optical sensor, along with an RF spectrum analyzer, were used to measure the system's average output power and assess the stability of the pulse repetition rate. The total cavity length, including the gain medium, is approximately 9.35 meters. The dispersion parameters of the EDF and the SMF are −17.3 and 17 ps/km·nm, respectively, resulting in a net dispersion of -0.12 ps^2 for the cavity.

In the anomalous dispersion regime, when the dispersion counterbalances the nonlinearity within the cavity, a mode-locked laser can be generated, either with conventional or bound solitons. The Kerr effect causes a nonlinear change in the refractive index, leading to NI-MMI of the excited modes in the GIMF. The SMS structure facilitates the self-imaging and spectral filtering effects, providing stable modulation to achieve nonlinear saturable absorption properties for light. Mode-locking operation is achieved at a wavelength where the SMF-GIMF-SMF structure acts as a saturable absorber. It is important to note that both Q-switching and mode-locking, using the SMS structure as an ASA, rely on the mechanism of nonlinear multimode

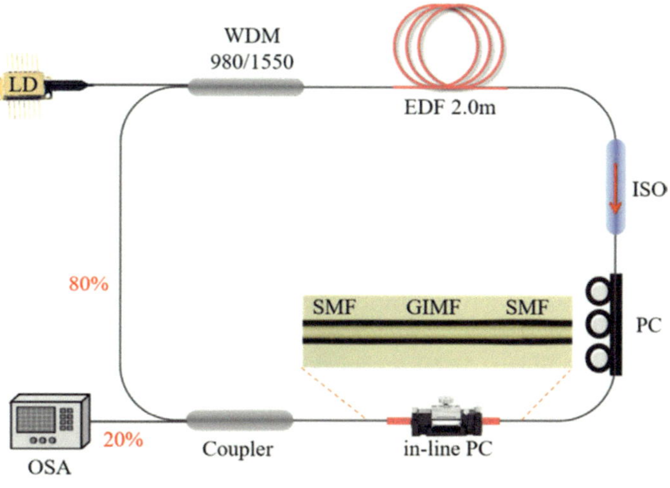

Figure 8.
Schematic diagram of the mode-locked fiber laser, with an inset showing a microscope image of the SMS structure.

interference. If light coupling is imperfect and the cavity experiences significant losses, both mode-locking and Q-switching can be achieved within the same cavity by adjusting the rotation angles of the waveplates or the polarization controller. In other words, mode-locking and Q-switching can coexist simultaneously in the fiber laser cavity with a fixed pump power. Additionally, in some cases, the transition between Q-switching and mode-locking can be accomplished simply by altering the pump power.

4.3 Generation of conventional soliton

As previously mentioned, mode excitation and coupling of the unidirectional oscillating photons at the joint of SMF and GIMF creates wavelength-dependent transmission. The degree of this multimode interference is heavily dependent on the effective refractive index within the fiber's core [38]. The Kerr effect, influencing the core refractive index with changing light intensity, facilitates high transmission at elevated power intensity and low transmission at lower intensity [39]. This behavior closely resembles that of an SA. In this scenario, the SMS structure is utilized as an ASA to modulate cavity loss for mode-locking. However, the SMS structure demonstrates sensitivity to the length of the GIMF. To address this, we introduced an in-line PC to alleviate the GIMF length limitation. This was achieved by applying pressure on the straight GIMF to alter its birefringence.

There are two ways to achieve pulse splitting with a constant pump power and cavity length: one is to reduce the modulation depth of the SA, and the other is to increase cavity dispersion [19]. We chose to reduce the modulation depth of SA. To explore this, we tested the modulation depths of the SMF-GIMF-SMF structure with different GIMF lengths (45, 100, 108, 156 mm). Shorter GIMF lengths resulted in lower modulation depths. Considering that shorter GIMF lengths also tolerate lower pump powers, we selected the 45 mm GIMF-based SMS structure for our demonstration. The experiment tested various GIMF lengths (45, 100, 108, 156 mm), all of which consistently exhibited stable mode-locking behavior. The results showed that longer GIMF lengths could handle higher power but had a lower pump power threshold for initiating self-started mode-locked pulses, with 156 mm GIMF reaching the lowest threshold at 17 mW. The mode-locked pulse generation stopped after 251 mW with the 45 mm GIMF, while the 156 mm GIMF demonstrated greater resilience, withstanding up to the maximum pump power of 343 mW. Importantly, the 45 mm GIMF-based SMS structure displayed distinctive bound soliton behavior, which made it the primary focus of our demonstration.

With proper adjustment of the PC, a stable mode-locked laser was observed on the oscilloscope at a pump power of 29.3 mW. The mode-locked laser's performance is summarized in **Figure 9**. **Figure 9(a)** shows a representative optical spectrum of the mode-locked pulses, which exhibits a conventional soliton spectrum with Kelly sidebands. Operating at a center wavelength of around 1573 nm, it has a 3 dB spectral bandwidth of 7.83 nm. **Figure 9(b)** presents the oscilloscope trace of the mode-locked pulse train, showing relatively uniform intensity with a pulse period of 44.34 ns, corresponding to the round-trip time within the cavity. The RF spectrum of the generated pulse train is displayed in **Figure 9(c)**, where a strong peak is observed at the cavity's fundamental repetition rate of 21.98 MHz, corresponding to a laser cavity length of 9.35 m. The SNR reached 68 dB, indicating highly stable mode-locking operation. The inset shows the 1-GHz span RF spectrum, suggesting that the fiber laser operates effectively in continuous mode-locking. After fitting with a $Sech^2$ curve, the autocorrelation trace in **Figure 9(d)** reveals a FWHM of 650 fs. The resulting time-bandwidth product (TBP) was calculated to be 0.615, exceeding the transform limit of 0.315, indicating that the pulse is up-chirped. This chirping is likely induced by the multimode fiber. To assess the long-term stability of the conventional soliton,

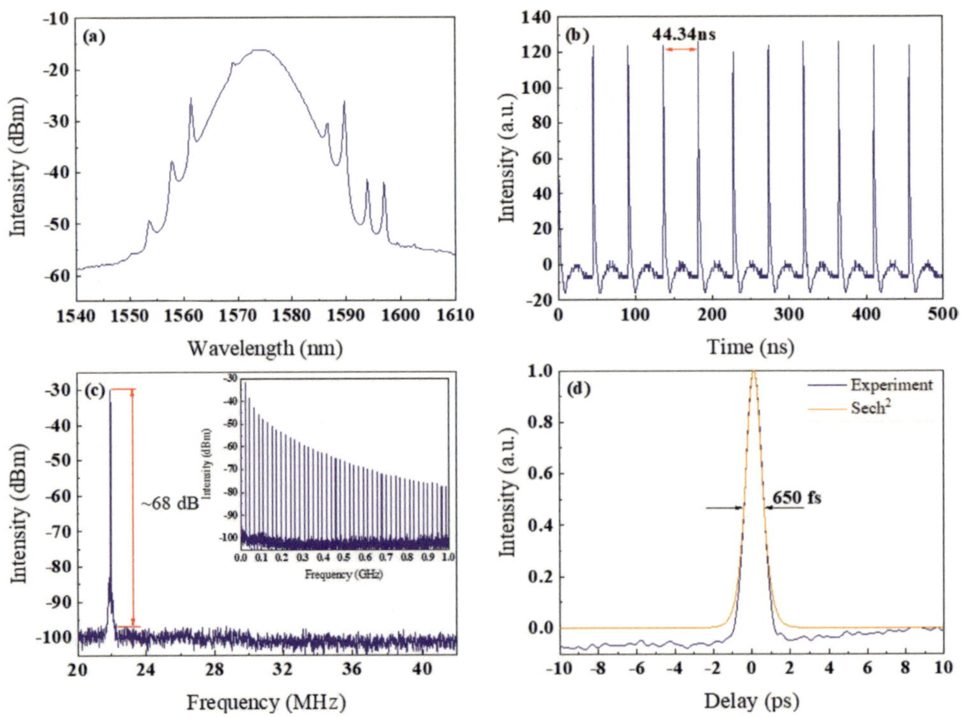

Figure 9.
Conventional soliton performances (a) Optical spectrum; (b) pulse train; (c) RF spectrum (inset: the RF spectrum within a larger range of 1 GHz); (d) autocorrelation trace.

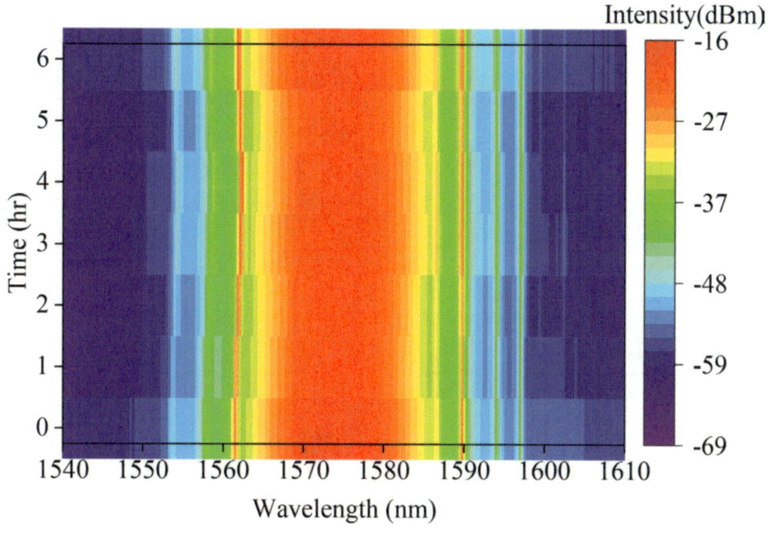

Figure 10.
Long-term stability test of the conventional soliton optical spectrum.

we recorded the optical spectrum every hour for 6 hours at a fixed pump power of 47.8 mW. The spectra remained consistent over time, as shown in **Figure 10**.

4.4 Generation of bound solitons

After observing the conventional soliton at the highest pump power it could withstand, 158.5 mW, we reduced the pump power to 54 mW. With proper adjustment of the PC, both tightly and loosely bound solitons were observed simultaneously. **Figure 11(a)** shows the optical spectrum, with an enlarged section highlighting a regular modulation with a period of approximately 0.6 nm. The center wavelength is around 1570 nm, and the FWHM is approximately 10 nm. **Figure 11(b)** displays the typical pulse train at a pump power of 54 mW, showing a pulse interval of 45.54 ns, which corresponds to a repetition rate of 21.98 MHz. Due to the limitations of the oscilloscope, we were unable to zoom in enough to observe the pulse splitting directly, but this can be indirectly seen in the autocorrelation trace. **Figure 11(c)** presents the corresponding RF spectrum, revealing multiple harmonics within a 500 MHz frequency range. An SNR of 37 dB is observed at the fundamental frequency of 21.98 MHz.

Figure 11(d) shows the autocorrelation trace, where a slight asymmetry is present due to measurement limitations. The three higher peaks, with an approximately 1:2:1 intensity ratio, indicate that the two solitons forming the bound-state pulse have nearly identical intensities. The time separation between them is around 12.8 ps, approximately 17.3 times the pulse width of 740 fs, suggesting that the two solitons are loosely bound. The two side peaks also have two smaller peaks around them, as one of the solitons in the loosely bound pair is itself a bound-state soliton [40]. The time separation of this bound-state soliton is 1.8 ps, about 2.4 times the pulse width, indicating that these solitons are tightly bound. The very short time separation between the solitons in the bound state suggests direct interaction, which may be due to attractive or repulsive forces acting on each soliton. The intensity ratio of the tightly bound soliton is approximately 1:3:1, likely because the intensities of the two

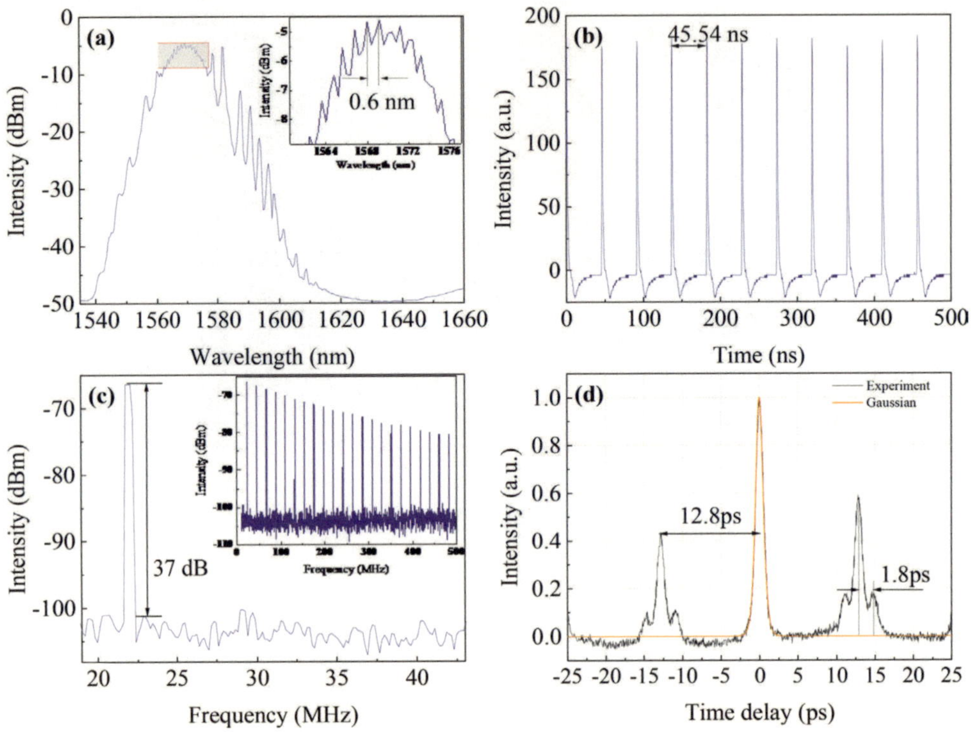

Figure 11.
Bound soliton performances (a) optical spectrum; (b) pulse train; (c) RF spectrum (inset: the RF spectrum within a range of 500 MHz); (d) autocorrelation trace.

solitons differ [41]. The formation of tightly and loosely bound solitons can also be attributed to non-Hermitian coupling, especially when cavity frequencies or subharmonic frequencies vary. This non-Hermitian coupling, characterized by distinct periodic frequencies, introduces time delays or frequency differences, which can lead to the spontaneous synchronization or localization of tightly bound solitons, as well as loosely bound ones. Additionally, this coupling can trigger subharmonic entrainment of pulses in ultrafast fiber lasers.

Due to the instability of the higher-order soliton, increasing the pump power caused the side peaks around the two higher side peaks to disappear. This is likely because the phase relationship between the two solitons weakened, causing the soliton to no longer be bound to the side soliton [42]. Once the side peaks disappeared, only three peaks remained in the autocorrelation trace. **Figure 12(a)-(c)** shows the autocorrelation trace, where the time separations remain at 20.4 ps. Initially, the intensity ratio of the peaks increases beyond 1:2:1, indicating that the intensities of the two solitons are no longer the same, and the difference between them becomes more pronounced. Over time, as the pump power increases, the intensity ratio of the peaks in the autocorrelation trace continues to rise. This is likely due to changes in reflectivity within the cavity caused by the increased pump power, which in turn alters the intensities of the solitons in the bound state. **Figure 12(d)-(f)** shows the optical spectrum, where the modulation depth decreases as the intensity ratio of the peaks increases, which aligns with theoretical predictions [43]. The modulation period in the spectrum remains nearly constant at 0.4 nm, and with a center wavelength of 1570 nm and a time separation of 20.4 ps, the observed phenomenon closely

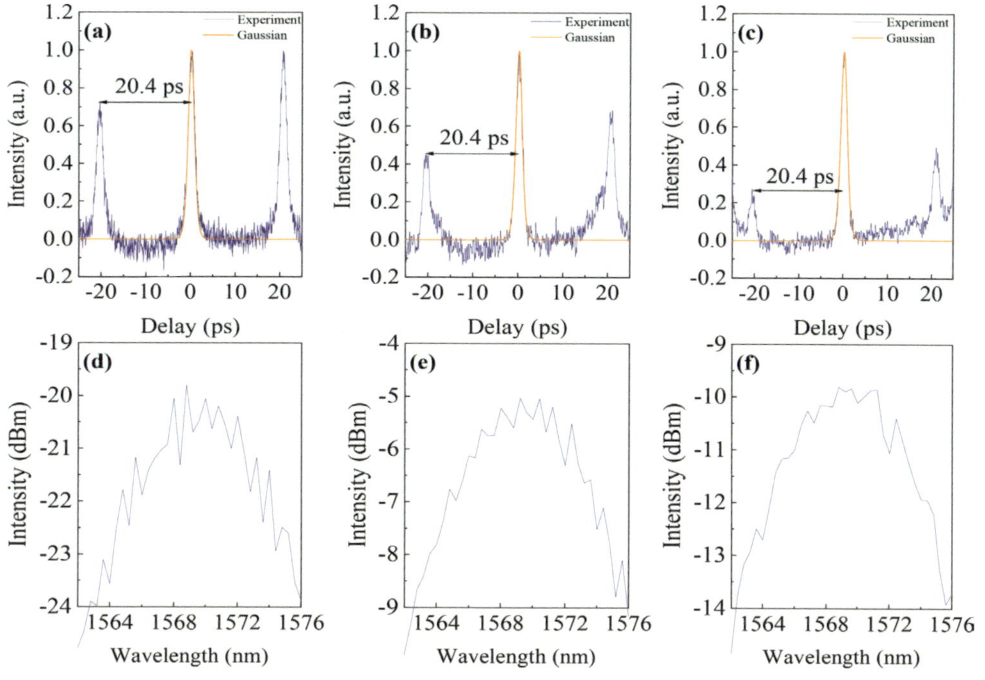

Figure 12.
Bound soliton characteristics at different pump power levels (a)-(c) Autocorrelator trace. (d)-(f) Optical spectrum.

matches the prediction from Eq. (8). Another intriguing observation is that, with the same cavity settings and the 45 mm SA, the conventional soliton disappears at around 140 mW in all our tests, whereas the pump power can reach up to 250.7 mW for bound soliton operation.

The results may inspire further exploration of bound states in optical telecommunication systems. Structured bound solitons provide a simple method to expand the coding alphabet, thereby boosting data transfer rates and increasing the capacity of fiber-optic networks [44]. Furthermore, a frequency comb can be generated via soliton mode-locking in a silica resonator [45]. By leveraging their nonlinear properties, soliton pulses enable the development of logic gates, memory, interconnects, and signal processing, laying the foundation for more efficient optical computing systems [46].

5. Conclusions

Q-switched and mode-locked EDFLs are widely used in fields such as telecommunications, bio-imaging, spectroscopy, atmospheric measurements, surgery, and material processing. This chapter highlights the potential of the single-mode multimode single-mode (SMS) structure as an effective approach for developing simple, compact, and cost-efficient Q-switched and mode-locked EDFLs. The SMS configuration acts as an innovative ASA, enabling the generation of short and ultrashort pulsed lasers through the Kerr effect and nonlinear multimode interference. By fusion-splicing single-mode fibers to both ends of a specific length of multimode fiber, the SMS structure is created. Integrating this SMS setup into a ring EDFL and fine-tuning its parameters with a polarization controller successfully demonstrated pulsed lasers, including bound solitons. A passively Q-switched EDFL was achieved using a 158 mm length of 105 µm core diameter SIMF to form the SMS structure as an ASA, modulating cavity loss for Q-switching through nonlinear multimode interference. At a maximum pump power of 343 mW, a pulse repetition rate of 157.7 kHz, pulse width of 2.5 µs, output power of 2.35 mW, and pulse energy of 15.0 nJ were recorded. Additionally, an SMF-GIMF-SMF structure was employed as an ASA to generate both conventional and bound solitons in a mode-locked fiber laser. The conventional soliton displayed exceptional long-term stability, operating at 1573 nm with a pulse width of 650 fs and a repetition rate of 21.98 MHz at a remarkably low pump power threshold of 29 mW. As pump power increased, bound solitons evolved from a mix of tightly and loosely bound solitons to purely loosely bound solitons. It was observed that, with varying pump power, loosely bound solitons maintained consistent time separations but exhibited differing peak ratios in the autocorrelation trace. The tightly and loosely bound solitons, centered around 1570 nm, had pulse widths of 740 fs with time separations of 12.8 and 1.8 ps, respectively. Moreover, several loosely bound solitons were identified with a fixed time separation of 20.4 ps but varying intensity ratios. These findings confirm the ability of both SMF-SIMF-SMF and SMF-GIMF-SMF structures to function as saturable absorbers, enabling the generation of stable Q-switched and mode-locked EDFLs with simple cavity designs.

Acknowledgements

The authors gratefully acknowledge the financial support from Airlangga University under grant No. 217/UN3.15/PT/2022 (Skema Hibah Riset Mandat Universitas Airlangga).

Conflict of interest

The authors declare no conflict of interest.

Author details

Sulaiman Wadi Harun[1]*, Yu Chen[2], Ahmad Haziq A. Rosol[1], Zian Cheak Tiu[3] and Retna Apsari[4]

1 Department of Electrical Engineering, University of Malaya, Kuala Lumpur, Malaysia

2 Chongqing Vocational Institute of Engineering, Chongqing, China

3 Faculty of Engineering and Quantity Surveying, INTI International University, Nilai, Negeri Sembilan, Malaysia

4 Department of Physics, Faculty of Science and Technology, Universitas Airlangga, Jl. Mulyorejo, Surabaya, Indonesia

*Address all correspondence to: swharun@um.edu.my

References

[1] Keller U. Recent developments in compact ultrafast lasers. Nature. 2003;**424**(6950):831-838

[2] Wadi NIS, Kadir RA, Jusoh Z, Ahmed N, Khudus MIM, Salam S. Utilizing iron phthalocyanines absorber for the generation of Q-switched and mode-locked pulses. Optik. 2024;**300**:171665

[3] Zhang Y et al. Study of the influence of SESAM parameters on the evolution of mode-locked pulses at different repetition rates. Applied Physics B. 2021;**127**(8):1-10

[4] Nishizawa N, Jin L, Kataura H, Sakakibara Y. Dynamics of a dispersion-managed passively mode-locked Er-doped fiber laser using single wall carbon nanotubes. Photonics. 2015;**2**(3: MDPI):808-824

[5] Ahmad H et al. S-band Q-switched fiber laser using molybdenum disulfide (MoS2) saturable absorber. Laser Physics Letters. 2016;**13**(3):035103

[6] Haris H, Harun SW, Yupapin P, Arof H, Apsari R. Generation of vector soliton pulses with graphene oxide film in mode-locked erbium-doped fiber laser cavity. Nonlinear Optics Quantum Optics-Concepts in Modern Optics. 2020;**52**(1-2):111-118. Available from: <Go to ISI>://WOS:000509998200006

[7] Hisyam MB, Rusdi MFM, Latiff AA, Harun SW. Generation of mode-locked ytterbium doped fiber ring laser using few-layer black phosphorus as a saturable absorber. IEEE Journal of Selected Topics in Quantum Electronics. 2016;**23**(1):39-43

[8] Chen Y, Cheak TZ, Jin TS, Vinitha G, Dimyati K, Harun SW. Domain-wall dark pulse generation with SMF-GIMF-SMF structure as artificial saturable absorber. Scientific Reports. 2024;**14**(1):2141

[9] Dong Z et al. Er-doped mode-locked fiber lasers based on nonlinear polarization rotation and nonlinear multimode interference. Optics and Laser Technology. 2020;**130**:106337

[10] Liu J et al. Duration switchable high-energy passively mode-locked Raman fiber laser based on nonlinear polarization evolution. IEEE Photonics Journal. 2015;**7**(5):1-7

[11] Ilday F, Wise F, Sosnowski T. High-energy femtosecond stretched-pulse fiber laser with a nonlinear optical loop mirror. Optics Letters. 2002;**27**(17):1531-1533

[12] Zhou R, Yu D, Liu X, Li Q, Fu H. Dark rectangular noise-like pulses in a figure-nine fiber laser based on a nonlinear amplifying loop mirror. Optics Letters. 2019;**44**(15):3717-3720

[13] Allison SW, Gillies GT. Observations of and applications for self-imaging in optical fibers. Applied Optics. 1994;**33**(10):1802-1805. DOI: 10.1364/AO.33.001802

[14] Zhu X. Multimode Interference in Optical Fibers and its Applications in Fiber Lasers and Amplifiers (Doctoral Dissertation). Tucson, United States: The University of Arizona; 2008. Available from: http://hdl.handle.net/10150/195318

[15] Nazemosadat E, Mafi A. Nonlinear multimodal interference and saturable absorption using a short graded-index multimode optical fiber. Journal of the Optical Society of America B. 2013;**30**(5):1357-1367. DOI: 10.1364/JOSAB.30.001357

[16] Fu S et al. Passive Q-switching of an all-fiber laser induced by the Kerr effect of multimode interference. Optics Express. 2015;**23**(13):17255-17262. DOI: 10.1364/OE.23.017255

[17] Chen Y, Dimyati K, Badruddin IA, Kamangar S, Harun SW. All-fiber Q-switched erbium-doped fiber laser generation with SMF-SIMF-SMF structure as artificial saturable absorber. Infrared Physics and Technology. 2024;**136**:105088

[18] Wang J et al. Wavelength and interval tunable second-order bound solitons in mode-locked fiber laser based on Lyot filter. Results in Physics. 2024;**58**:107479. DOI: 10.1016/j.rinp.2024.107479

[19] Zhu T, Wang Z, Wang DN, Yang F, Li L. Observation of controllable tightly and loosely bound solitons with an all-fiber saturable absorber. Photonics Research. 2019;**7**(1):61-68. DOI: 10.1364/PRJ.7.000061

[20] Chen J, Wang DN, Wang Z. Wavelength-switchable multiple type bound solitons in a passively mode-locked Er-doped fiber laser. IEEE Photonics Technology Letters. 2020;**32**(22):1447-1450. DOI: 10.1109/LPT.2020.3031919

[21] Demircan A, Bandelow U, Pietrzyk M, Kanatchikov I. High-Order solitons and modulation instability in optical fibers. In Book: Handbook of Solitons: Research, Technology and Applications. New York: Nova Science Publishers, Inc.; 2009

[22] Malomed BA. Bound solitons in coupled nonlinear Schrödinger equations. Physical Review A. 1992;**45**(12):R8321

[23] Akhmediev N, Ankiewicz A, Soto-Crespo J. Multisoliton solutions of the complex Ginzburg-Landau equation. Physical Review Letters. 1997;**79**(21):4047

[24] Tang D, Zhao L, Zhao B, Liu A-Q. Mechanism of multisoliton formation and soliton energy quantization in passively mode-locked fiber lasers. Physical Review A. 2009;**72**:043816. DOI: 10.1103/PhysRevA.72.043816

[25] Sakaguchi H, Skryabin DV, Malomed BA. Stationary and oscillatory bound states of dissipative solitons created by third-order dispersion. Optics Letters. 2018;**43**(11):2688-2691. DOI: 10.1364/OL.43.002688

[26] Kulagin AE, Shapovalov AV. A semiclassical approach to the nonlocal nonlinear Schrödinger equation with a non-Hermitian term. Mathematics. 2024;**12**:580. DOI: 10.3390/math12040580

[27] Tang D, Zhao L, Zhao B. Multipulse bound solitons with fixed pulse separations formed by direct soliton interaction. Applied Physics B. 2005;**80**(2):239-242

[28] Seong N, Kim DY. Formation of bound solitons and their properties in a figure-eight fiber laser. In: Optical Fiber Communication Conference. Anaheim, United States: Optica Publishing Group; 2002. p. ThGG19

[29] Pang M, He W, Jiang X, Russell PSJ. All-optical bit storage in a fibre laser by optomechanically bound states of solitons. Nature Photonics. 2016;**10**(7):454-458

[30] He W, Pang M, Yeh D-H, Huang J, Russell PSJ. Synthesis and dissociation of soliton molecules in parallel optical-soliton reactors. Light: Science and Applications. 2021;**10**(1):120

[31] Wang Z et al. Generation of pulse-width controllable dissipative solitons

and bound solitons by using an all fiber saturable absorber. Optics Letters. 2019;**44**(3):570-573

[32] Ren X, Li H, Zhang Z, Zhang J, Xu S. Generation of bound solitons in a mode-locked Tm-doped fiber laser based on a graded-index multimode fiber saturable absorber. Optics and Laser Technology. 2022;**149**:107865. DOI: 10.1016/j.optlastec.2022.107865

[33] Zhang H et al. All-fiber nonlinear optical switch based on polarization controller coiled SMF–GIMF–SMF for ultrashort pulse generation. Optics Communications. 2019;**452**:7-11. DOI: 10.1016/j.optcom.2019.07.004

[34] Chen J et al. GIMF-based SA for generation of high pulse energy ultrafast solitons in a mode-locked linear-cavity fiber laser. Journal of Lightwave Technology. 2020;**38**(6):1480-1485. DOI: 10.1109/jlt.2019.2954828

[35] Gan Y et al. Generation of high-order solitons with order continuously adjustable in a fiber laser based on GIMF–SIMF–GIMF saturable absorber. Optics Communications. 2021;**479**:126441. DOI: 10.1016/j.optcom.2020.126441

[36] Qi Y et al. Tunable all fiber multi-wavelength mode-locked laser with a large dynamic range using polarization controller coiled SMF-GIMF-SMF structure as both saturable absorber and comb filter. Optical Fiber Technology. 2022;**74**:103055. DOI: 10.1016/j.yofte.2022.103055

[37] Yang Y, Jiang F, Zeng Q, Wang K, Wang Z, Song Y. Thulium-doped multi-wavelength and mode-locked fiber laser based on multimode interference. Infrared Physics and Technology. 2023;**133**:104869. DOI: 10.1016/j.infrared.2023.104869

[38] Fu S et al. Passive Q-switching of an all-fiber laser induced by the Kerr effect of multimode interference. Optics Express. 2015;**23**(13):17255-17262

[39] Zhao W, Chen G, Li W, Wang G, Zeng C. All-fiber saturable absorbers for ultrafast fiber lasers. IEEE Photonics Journal. 2019;**11**(5):1-19. DOI: 10.1109/jphot.2019.2941580

[40] Li W et al. Generation of separation-locked bound solitons in a passively mode-locked all-fiber laser with a Fabry-Perot microcavity. Optics and Laser Technology. 2022;**150**:107936. DOI: 10.1016/j.optlastec.2022.107936

[41] Song YF, Zhang H, Zhao LM, Shen DY, Tang DY. Coexistence and interaction of vector and bound vector solitons in a dispersion-managed fiber laser mode locked by graphene. Optics Express. 2016;**24**(2):1814-1822. DOI: 10.1364/OE.24.001814

[42] Grapinet M, Grelu P. Vibrating soliton pairs in a mode-locked laser cavity. Optics Letters. 2006;**31**(14):2115-2117. DOI: 10.1364/OL.31.002115

[43] Luo AP et al. Observation of three bound states from a topological insulator mode-locked soliton fiber laser. IEEE Photonics Journal. 2014;**6**(4):1-8. DOI: 10.1109/JPHOT.2014.2345874

[44] Wang P, Zhao K, Gui L, Xiao X, Yang C. Self-organized structures of soliton molecules in 2-$\mu $ m fiber laser based on MoS 2 saturable absorber. IEEE Photonics Technology Letters. 2018;**30**(13):1210-1213

[45] Yi X, Yang Q-F, Yang KY, Suh M-G, Vahala K. Soliton frequency

comb at microwave rates in a high-Q silica microresonator. Optica. 2015;**2**(12):1078-1085

[46] Ghadi A. All-optical computing circuits half-subtractor and comparator based on soliton interactions. Optik. 2021;**227**:166079

Chapter 6

Spectral Sidebands Formed in Soliton Fiber Lasers

Hu Xiao and Dingyuan Tang

Abstract

Spectral sideband formation is a fundamental phenomenon in soliton fiber lasers, arising from multiple mechanisms. The most recognized are Kelly sidebands, generated from the linear interference between a soliton and dispersive waves due to periodic perturbations in the cavity. Modulation instability (MI) is another key driver, creating sidebands in both anomalous and normal dispersion regimes through conventional and cavity-aided MI. Furthermore, forced periodic variations in soliton parameters like energy or pulse width can induce additional sidebands, while nonlinear interactions between orthogonally polarized components generate polarization instability-induced sidebands. Studying these sidebands provides critical insight into soliton dynamics, formation, and stability, which is essential for optimizing ultrafast lasers for telecommunications and high-power applications. This examination covers the properties of these various sidebands and discusses their significant influence on soliton operation. The text begins with a theoretical description of sidebands predicted for scalar and coupled nonlinear Schrödinger equations, summarizes key research results, and concludes with a discussion on future research directions.

Keywords: fiber lasers, nonlinear optics, optical solitons, soliton dynamics, spectral sidebands

1. Introduction

Spectral sideband formation in single-mode fiber (SMF) lasers is a feature that has attracted considerable attention of research. It could be caused by various effects such as nonlinear wave interactions, modulation instability (MI), and soliton dynamics. These spectral sidebands provide valuable insights into the underlying physics of SMF laser and the evolution of solitons governed by the nonlinear Schrödinger equation (NLSE) and its coupled versions [1–6].

In the context of the scalar NLSE, modulation instability (MI) plays a critical role in the formation of spectral sidebands. MI arises due to the interplay between group velocity dispersion (GVD) and self-phase modulation (SPM), leading to the exponential amplification of spectral perturbations [7, 8]. Furthermore, in SMF lasers, periodic boundary condition modifies MI features, leading to cavity-induced MI (CIMI) sidebands, where the CIMI could occur in both the normal and the anomalous dispersion regimes [9, 10]. In addition to these MI-induced sidebands, traditional Kelly soliton sidebands can also appear [11, 12], often as a result of periodic perturbations and linear interference between the perturbed soliton and dispersive waves.

In the vector (coupled) NLSE framework, polarization modulation instability (PMI) sidebands emerge due to cross-phase modulation (XPM) between the two orthogonal polarization components [13, 14]. PMI sidebands often appear at different frequencies from the MI sidebands, influenced by birefringence and polarization mode dispersion. Additionally, the complexity of coupled NLSE may lead to emergence of new soliton sidebands. These include group velocity-locked soliton sidebands [15], where two solitons move together and interact incoherently. Furthermore, four-wave mixing (FWM)-induced soliton sidebands arise from the coherent energy exchange between vector solitons [16]. These nonlinear interactions play a crucial role in shaping the spectral structure of soliton fiber lasers and significantly impact their stability. This chapter provides a comprehensive theoretical study of the different types of sidebands formed in SMF lasers.

2. Theoretical studies

2.1 Pulse propagation in scalar SMFs

When a high-intensity light beam travels through a single-mode fiber (SMF), its propagation dynamics are influenced by the interplay between dispersion and nonlinear effects. Dispersion, an inherent material property, arises from the frequency-dependent variation of the group velocity. In SMFs, pulse broadening primarily occurs due to chromatic dispersion and polarization mode dispersion. Alongside dispersion, nonlinear effects manifest as intensity-dependent changes in the phase velocity of the pulse. Among these, the optical Kerr effect is particularly significant, inducing self-phase modulation (SPM) in the propagating light.

Mathematically, nonlinear pulse propagation in a scalar SMF is described by the following NLSE [12]:

$$\frac{\partial A}{\partial z}=-i\frac{\beta_2}{2}\frac{\partial^2 A}{\partial t^2}+i\gamma|A|^2A \tag{1}$$

Where A is the slowly varying pulse envelope of the optical pulse. β2 is the second-order dispersion coefficient, and γ is the nonlinear coefficient of the fiber. In nonlinear optics, the NLSE governs a wide range of nonlinear optical phenomena in fibers, including various types of modulation instability (MI) effects, and soliton solutions such as bright and dark solitons.

2.1.1 Modulation instability in SMFs

Modulation instability is a process in which the amplitude and phase modulation of a wave grow as a result of an interplay between the nonlinearity and anomalous dispersion. It has been shown that the CW solution $\overline{A}=\sqrt{P_0}\exp(i\phi_{NL})$ [3] of Eq. (1) is unstable for a small perturbation (or modulation) $a(z,t)=a_1\exp[i(Kz-\Omega T)]+a_2\exp[-i(Kz-\Omega T)]$, where P_0 is the incident power and ϕ_{NL} is the nonlinear phase shift induced by SPM. $K$and Ω are the wave number and frequency of perturbations satisfying the following relation [3],

$$
\begin{aligned}
g(\Omega) &= \Omega_{\max}\left(\Omega_c^2 - \Omega^2\right)^{1/2} \\
\Omega_{\max} &= \pm\frac{\Omega_c}{\sqrt{2}} = \pm\left(\frac{2\gamma P_0}{|\beta_2|}\right)^{1/2} \\
g_{\max} &\equiv g(\Omega_{\max}) = \frac{1}{2}|\beta_2|\Omega_c^2 = 2\gamma P_0
\end{aligned}
\tag{2}
$$

When the peak power of the pulse P_0 and Ω_c represents the cutoff frequency of the modulation instability (MI) gain spectrum, the dispersion relation reveals that the stability of the steady state is strongly influenced by the group-velocity dispersion (GVD) regime of the fiber. Under normal GVD ($\beta_2 > 0$), the wavenumber K remains real for all frequencies Ω, ensuring stability against small disturbances. However, in the anomalous GVD regime ($\beta_2 < 0$), K turns imaginary for frequencies, causing perturbations $a(z, t)$ to amplify exponentially along the propagation distance z. This phenomenon, known as modulation instability, induces temporal and spectral reshaping of a continuous-wave (CW) beam, with the MI gain determining the extent of these modulations is given by,

$$g(\Omega) = |\beta_2\Omega|\left(\Omega_c^2 - \Omega^2\right)^{1/2} \tag{3}$$

Where the maximum gain frequency is given by,

$$\Omega_{\max} = \pm\frac{\Omega_c}{\sqrt{2}} = \pm\left(\frac{2\gamma P_0}{|\beta_2|}\right)^{1/2} \tag{4}$$

with a peak value as shwon in **Figure 1**,

$$g_{\max} \equiv g(\Omega_{\max}) = \frac{1}{2}|\beta_2|\Omega_c^2 = 2\gamma P_0 \tag{5}$$

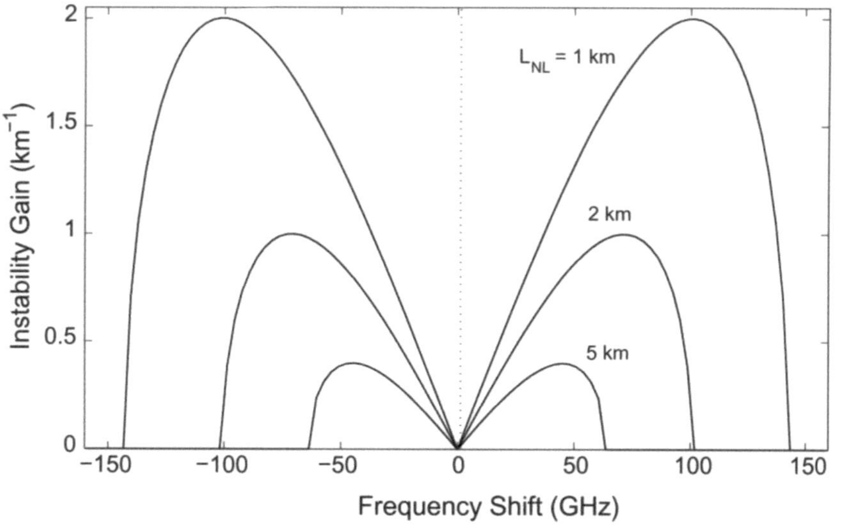

Figure 1.
Gain spectra of modulation instability for three values of the nonlinear length [©Agrawal et al. Nonlinear Fiber Optics Sixth edition].

Modulation Instability (MI) is a parametric process in nonlinear optical systems, where small perturbations on a continuous wave (CW) are amplified through a Four-Wave Mixing (FWM) process between the CW wave and the modulation or perturbation wave. This amplification results in the generation of symmetrical sidebands around the carrier frequency. The amplitude of the sidebands is proportional to A_1^n, where A_1 is the amplitude of the initial perturbation on the optical wave [17, 18], and is separated from the carrier frequency by $\frac{n\Omega_c}{\sqrt{2}}$, $n=1, 2 \ldots$.

2.1.2 Periodic power variation-induced sideband instability in a SMF laser

It has been shown by Matera et al. that periodic light power variation could lead to formation of a parametric sideband as a result of parametric coupling between the light and the background noise [19]. As light propagates within the fiber cavity, a fraction of its power is extracted at the output coupler with each round trip, leading to an abrupt decrease in the intracavity power. Under steady-state laser operation, the power drop is compensated by amplification of the gain when it propagates in the gain fiber. To consider the periodic light power variation, Eq. (1) becomes,

$$\frac{\partial A}{\partial z} = -i\frac{\beta_2}{2}\frac{\partial^2 A}{\partial t^2} + i\gamma q(z)|A|^2 A \tag{6}$$

Where $q(z) = \exp(2gz)$ is a periodic function of the cavity round trips and g is the averaged laser gain coefficient along the fiber. According to reference [19], parametric instability would arise in a system when wavevector of $q(z)$ is twice of the wave vector of the light field, and the positions of the sidebands are at [19]:

$$\Omega_{peak} = \pm\left(\frac{2\pi p}{\beta_2 l} - \frac{2\gamma P_0 c_0}{\beta_2}\right)^{1/2} \tag{7}$$

$P_0 c_0$ denotes the average power in the fiber ring cavity, where l represents the cavity length and p is the order in the Fourier expansion of the perturbation $q(z)$. The sideband instability arises from a four-wave mixing (FWM) process, facilitated by the periodic modulation of optical power within the cavity, which acts as an effective grating. This instability occurs when the below phase-matching condition is satisfied,

$$k_1 + k_2 = k_{grating} + 2k_{pump} \tag{8}$$

Where k_1 and k_2 are the wavevector of the two induced parametric sidebands and $k_{grating}$ and k_{pump} are the wavevector of the artificial grating and the laser beam.

2.1.3 Cavity-induced modulation instability in a SMF laser

Theoretical studies have shown that traditional MI (described in Section 1.1.1) can occur only in anomalous fiber dispersion regime and there is no MI for light propagation in normal dispersion fibers. However, Haelterman et al. have theoretically shown that MI could still occur in a normal dispersion fiber cavity [20, 21]. Studies showed that the cavity's boundary conditions allow phase synchronization between a high-power intracavity laser field and a pair of weak sidebands, resulting in efficient energy exchange. Referred to as cavity-induced modulation instability (CIMI), this effect contrasts with the traditional MI by being dispersion-sign-independent. While conventional MI requires anomalous dispersion, CIMI emerges irrespective of whether the cavity exhibits normal or anomalous dispersion. Tang et al. further extended the theory of CIMI to the case of fiber lasers, and the light propagation in a fiber ring cavity is described by the extend NLSE of the form [22],

$$\frac{\partial E^n}{\partial z} = -i\frac{\beta_2}{2}\frac{\partial^2 E^n}{\partial t^2} + i\gamma|E^n|^2E^n + gE^n \tag{9}$$

The amplitude of the optical field in the n-th fiber cavity round trip is denoted as E^n; the gain coefficient of the gain fiber is represented by g. z corresponds to the light propagation distance, and t is the retard time in a reference moving at the speed of light. For lasing operation in an active fiber ring resonator, the circulating light must satisfy the resonant cavity boundary conditions.

$$E^{n+1}(0,t) = \sqrt{\rho}E^n(L,t)\exp(i\psi_0) \tag{10}$$

L denotes the cavity length; ρ is the light intensity transmission rate of the output coupler in the fiber ring laser. $\psi_0$is the linear phase delay generated after propagating one cavity round trip, which is related to the optical path length of the cavity roundtrip.

Eq. (9) can be simplified to dimensionless with the following normalization: $\xi = z/L, U^n = \sqrt{\gamma L}E^n, \tau = t/\sqrt{|\beta_2|L}, G = gL, \eta = sgn\ (\beta_2)$ and becomes,

$$\frac{\partial U^n}{\partial \xi} = -i\frac{\eta}{2}\frac{\partial^2 U^n}{\partial \tau^2} + i|U^n|^2U^n + GU^n \tag{11}$$

Eq. (9) has the following CW solution:

$$U^n = \left(U_0e^{G\xi}\right)\exp\left(i\frac{U_0^2}{2G}\exp(2G\xi)\right) \tag{12}$$

Where U_0 is the light intensity at the beginning of the n-th cavity round trip. Following the same procedure in Ref. [3], the authors had analyzed the stability of CW solution under weak sideband perturbations, to this end, also add perturbation terms v to the CW solution in (12),

$$U^n = \left(U_0e^{G\xi} + v\right)\exp\left(i\frac{U_0^2}{2G}\exp(2G\xi)\right) \tag{13}$$

Taking Eq. (11) back into Eq. (6.9) and solve for v, the eigenvalue is $\mu = \Omega\sqrt{-\frac{\Omega^2}{4} - \eta U_0^2}$. Where the eigenvalues of the matrix in Ref. [22] are $q_\pm = \sqrt{\rho}\left(p \pm \sqrt{p^2 - 1}\right)$. Where $p = cos\ (\phi)\ cos\,h(\mu) - \frac{\eta\Omega^2/2 + U_0^2}{\mu} sin\ (\phi)\ sin\,h(\mu)$. It has been shown that CIMI will occur if the modulus of the eigenvalues is larger than unity. And the gain coefficient is defined by $G_{MI} = |q_\pm| - 1$, and the gain spectrum is determined by the total phase delay ϕ, as shown in **Figure 2**. The cavity phase detuning is expressed as δ = φ - 2mπ, where m represents an integer. When the laser frequency matches a cavity resonance (δ = 0), light propagation remains unaffected by the cavity. However, CIMI emerges whenever the accumulated phase shift deviates from 2mπ. Notably, CIMI exhibits two key properties: Its modulation frequency depends on

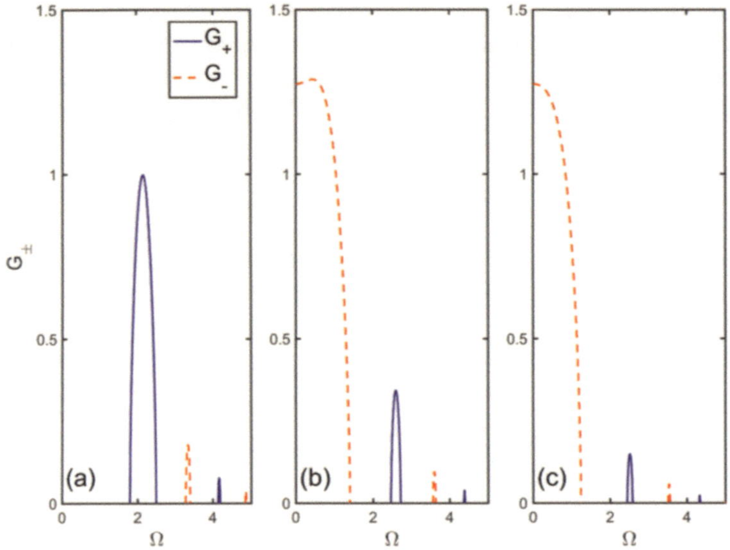

Figure 2.
Simulated gain spectrum of CIMI for both the normal and anomalous dispersion cases under two different linear phase delays.

the detuning value, and it can develop under both normal and anomalous dispersion conditions.

2.1.4 Soliton sidebands

Soliton sideband, also known as Kelly sideband, is the widely known spectral sidebands [23]. Different from the parametric sidebands mentioned in Section 1.1.1 to 1.1.3, which arise from the nonlinear four-wave mixing (FWM) process between a continuous-wave (CW) laser beam and perturbations where their onset requires sufficiently strong light intensity. Kelly sidebands arise from a linear interference process involving periodic perturbations of solitons and dispersive waves within a laser cavity. As a soliton propagates in a ring fiber laser, it inevitably encounters disturbances from intracavity elements, which can trigger the emission of dispersive waves. When the soliton completes one round trip and the phase difference between it and the emitted dispersive waves equals an integer multiple of 2π, constructive interference occurs. This results in the appearance of sharp spectral peaks, commonly referred to as Kelly sidebands. Their presence is a hallmark of soliton operation and is widely observed in soliton-based fiber lasers. Kelly et al. have first theoretically predicted the formation of the spectral sidebands based on the dimensionless NLSE with a constant term Γ_g,

$$-i\frac{\partial E}{\partial z}+\frac{1}{2}\frac{\partial^2 E}{\partial t^2}+|E|^2E=-i\Gamma_g E \tag{14}$$

Subject to a periodic loss provided by,

$$E(t,z=0)=\sqrt{1/G_s}\cdot E(t,z=z_a) \tag{15}$$

Where z_a is the amplification period and $G_s = \exp\left(+2\Gamma_g z_a\right)$ is the single pass gain. It has been shown that the constructive interference of the dispersive waves happens when soliton wavevector satisfies $k{=}2\pi n/z_a$. In terms of frequency, the frequency separation from the soliton carrier frequency δv_n follows the relation,

$$\delta v_n = \pm \frac{1}{2\pi t_p} \sqrt{1 + 8nZ_0/Z_a} \tag{16}$$

$\tau{=}2\ln\left[1{+}\sqrt{2}\right]\bullet t_p$ is the pulse's full pulse width at half maximum and $Z_{0,}$ and Z_a are the actual soliton and amplifier periods, respectively.

2.2 Pulse propagation in linearly birefringent SMFs

When an optical pulse propagates in a linearly birefringent fiber, the presence of two orthogonal polarization components leads to nonlinear interactions that make the pulse evolution far more intricate than in the scalar NLSE case. Mathematically, nonlinear pulse propagation in a linear birefringent SMF is described by the following coupled NLSEs [7],

$$\begin{aligned} \frac{\partial \mathrm{u}}{\partial \mathrm{z}} &= -\mathrm{i}\beta \mathrm{u} - \delta \frac{\partial \mathrm{u}}{\partial \mathrm{t}} - \mathrm{i}\frac{\beta_2}{2}\frac{\partial^2 u}{\partial \mathrm{t}^2} + \mathrm{i}\gamma\left(|u|^2 + \frac{2}{3}|v|^2\right)\mathrm{u} + \mathrm{i}\frac{\gamma}{3}v^2 u^* \\ \frac{\partial \mathrm{v}}{\partial \mathrm{z}} &= -\mathrm{i}\beta \mathrm{v} + \delta \frac{\partial \mathrm{v}}{\partial \mathrm{t}} - \mathrm{i}\frac{\beta_2}{2}\frac{\partial^2 v}{\partial \mathrm{t}^2} + \mathrm{i}\gamma\left(|v|^2 + \frac{2}{3}|u|^2\right)\mathrm{v} + \mathrm{i}\frac{\gamma}{3}u^2 v^* \end{aligned} \tag{17}$$

Here, u and v denote normalized slowly varying pulse envelopes along the slow and fast axes. u^* and v^* are the conjugates of u and v. $2\beta{=}2\Delta n/\lambda$ describes the wave number difference between the two polarization modes, where Δn represents the refractive index difference between the two polarization axes, $2\delta{=}2\beta\lambda/2\pi c$ is the group velocity mismatch. β2 corresponds to the second-order dispersion coefficient, and γ is the nonlinear coefficient of the fiber. The equations are written in a coordinate system that moves with the averaged group velocity.

Unlike the scalar nonlinear Schrödinger equation (NLSE), the coupled NLSE (CNLSE) exhibits enhanced complexity in nonlinear interactions, primarily through cross-phase modulation (XPM) and four-wave mixing (FWM)—represented by the fourth and final terms in Eq. (17), respectively. These additional effects give rise to more sophisticated soliton behavior, polarization-dependent phenomena, and energy transfer between interacting modes.

The nature of coupling between polarization modes depends on the fiber's birefringence strength. In single-mode fibers (SMFs) with negligible birefringence, coherent cross-polarization coupling (XPC) dominates, where phase matching between orthogonal polarization modes is inherently satisfied. Under these conditions, the group velocity mismatch term (δ) in Eq. (17) becomes insignificant. Conversely, incoherent XPC occurs when the phase difference between modes fluctuates rapidly compared to the nonlinear medium's response time. In this regime, refractive index modulation induced by one beam affects the other, governed primarily by their relative intensities rather than phase matching. As a result, the FWM contribution in Eq. (17) can be disregarded.

2.2.1 Modulation instability in birefringent SMFs

When multiple optical fields travel together through an optical fiber, they interact *via* the fiber's nonlinear effects. The most prevalent form of this interaction is cross-phase modulation (XPM). Studies have demonstrated that XPM between co-propagating light waves in single-mode fibers (SMFs) can give rise to a variety of novel optical phenomena. In 1987, Agrawal theoretically predicted that modulation instability (MI) could occur in normal dispersion fiber through incoherent coupling between two or more optical fields, even in cases where scalar MI described in section A.1 is not possible [24]. The gain of the incoherent coupling-induced MI is given by,

$$g(\Omega) = \beta_j|\Omega|\left(\Omega_c^2 - \Omega^2\right)^{1/2} \tag{18}$$

Where $\Omega_c^2{=}1/2\{[(b_1{+}b_2)^2{+}12b_1b_2]^{1/2}{-}(b_1{+}b_2)\}$ and $b_j{=}\left(4\gamma_j/\beta_j\right)P_j$. β_j is the second-order dispersion coefficient. $j{=}1,2$ accounts for the two polarization modes. $P_j$represents the optical power of the two polarization modes. Further, the wavenumber K and frequency of modulation Ω must satisfy the dispersion relation,

$$\left[(K - \Omega/v_{g1})^2 - f_1\right]\left[(K - \Omega/v_{g2})^2 - f_2\right] = C^2 \tag{19}$$

Where$f_j{=}\frac{1}{2}\beta_j\Omega^2\left(\frac{1}{2}\beta_j\Omega^2{+}2\gamma_jP_j\right)$, v_{gj} is the group velocity. And the coupling parameter C is given by $C{=}2\Omega^2(\beta_1\beta_2\gamma_1\gamma_2P_1P_2)^{1/2}$. Clearly, in the absence of XPC, C = 0, and Eq. (17) has the solution $K{=}\Omega/v_{gj}{\pm}\sqrt{f_j}$for $j{=}1,2$. This is the dispersion relation obtained in Section 1.1.1 for the single field case where in the normal dispersion regime, $\beta_j{>}0$, it follows that$f_j > 0$. Consequently, the parameter K remains real, preventing the occurrence of modulation instability (MI) in this regime. However, the scenario changes significantly when cross-phase modulation (XPM) introduces coupling between two optical fields. As shown in Eq. (19), K can become complex for certain values of Ω, even if both β_1 and β_2 are positive. Further, the gain of the MI relates not only to the power ratio between the two optical fields but also to the group velocity mismatch $\delta{=}\left|v_{g1}^{-1} - v_{g2}^{-1}\right|$ by Eq. (19). **Figure 3a** shows the gain spectra for several power levels P_1 and P_2 of the two beams and **Figure 3b** shows the gain spectra for several values of δ by use of Eqs. (18) and (19).

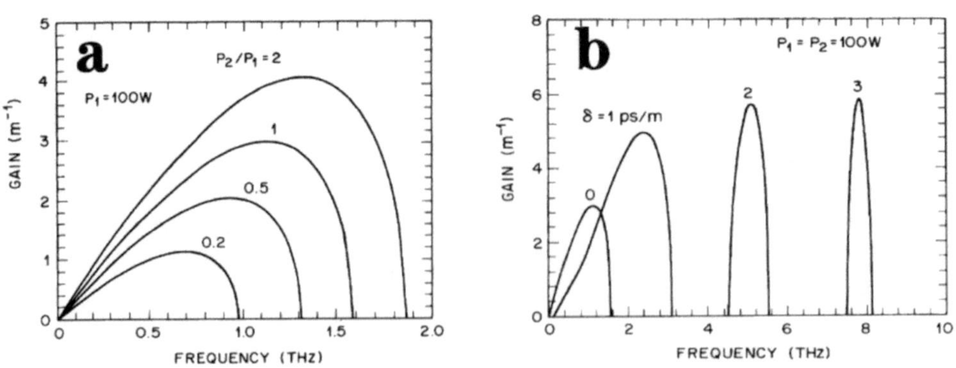

Figure 3.
(a)Gain spectra for several power levels of the two beams. (b) Effect of group-velocity mismatch δ in the gain spectra for the case $P_1{=}P_2{=}100W$ [© Ref. [24]].

2.2.2 Polarization modulation instability in birefringent SMFs

In 1988, Wabnitz provided a theoretical prediction of a type of polarization modulation instability using the coupled nonlinear Schrödinger equations. He demonstrated that coherent cross-phase modulation between two circularly polarized modes in a weakly birefringent single-mode fiber (SMF) can give rise to modulation instability in both normal and anomalous dispersion regimes. In particular, unlike conventional MI or MI that occurred under incoherent XPC, where the MI gain peaks were always located at a very high modulation frequency, under coherent cross-polarization coupling, the modulation instability gain peaks could be situated at a low or even vanishing modulation frequency [13, 25, 26]. Later, Chiu and Chow theoretically demonstrated that PMI could also arise from the coupling between two orthogonal linearly polarized modes. They further showed that a finite birefringence in the fiber could enhance PMI, with its characteristics governed by Eq. (15) [14]. In their work, the perturbations are introduced and modal dependence of $\exp[i(Kx - \Omega t)]$is sought in the standard manner, and the solutions for the perturbation frequency can be expressed as $\Omega^2 = \xi_1 \pm \sqrt{\xi_2}$, the parameters ξ_1 and ξ_2 are given by,

$$
\begin{aligned}
\xi_1 &= K^4\eta^2 + K^2\eta(\lambda - 1)\left(a^2 + b^2\right) + 4a^2b^2\lambda(\lambda + \sigma - 1) \\
\xi_2 &= \left[K^4\eta^2 + K^2\eta(\lambda - 1)\left(a^2 + b^2\right) + 4a^2b^2\lambda(\lambda + \sigma - 1)\right]^2 \\
&+K^2\eta\left[K^2\eta + 2\lambda\left(a^2 + b^2\right)\right]\left[-K^4\eta^2 + 2K^2\eta\left(a^2 + b^2\right) + 4a^2b^2\left(\lambda^2 + 2\sigma\lambda + \sigma^2 - 1\right)\right]
\end{aligned} \tag{20}
$$

Where K is the wavenumber, $\eta = \pm 1/2$ is the normalized second-order dispersion coefficient. λ is the normalized FWM coefficient in Eq. (17). Where nonzero λ indicates the presence of phase coherence nonlinearity. σ is the XPC coefficient, for orthogonal linearly polarized modes, σ=2/3. It has been shown that the two parameters ξ_1 and ξ_2 are real but the instability will arise if Ω is complex. Further, the presence of phase-sensitive coherently coupled nonlinearity, measured by λ, will enhance this MI. **Figure 4** shows the simulated results on the gain of PMI with the presence of λ.

2.2.3 Group-velocity locked vector solitons in SMFs

When a single-mode fiber (SMF) exhibits strong linear birefringence, the four-wave mixing (FWM) term in Eq. (17) becomes negligible. As a result, the interaction between the two polarization components becomes incoherent. Using the incoherently coupled nonlinear Schrödinger equations, Menyuk was the first to numerically investigate vector bright soliton solutions [27]. His findings revealed that for solitons with low amplitude, birefringence causes the two orthogonally polarized solitons to drift apart. However, if the soliton amplitude exceeds a certain threshold, the solitons can shift their frequencies in opposite directions to counteract the effects of birefringence.

Incoherently coupled bright solitons are also referred to as group velocity locked (GVL) bright solitons. Further theoretical investigations have revealed that, in addition to GVL bright solitons, other forms such as GVL dark solitons and GVL dark-bright solitons can also arise in birefringent single-mode fibers (SMFs). These solitons form through incoherent cross-polarization coupling (XPC) between orthogonally polarized modes. When two solitons with slightly differing group velocities interact, they can mutually trap each other by dynamically adjusting their central wavelengths

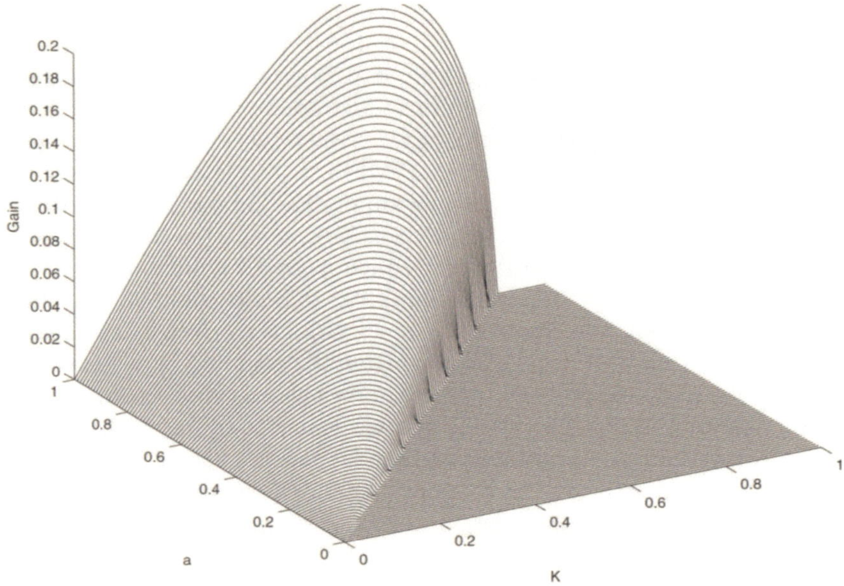

Figure 4.
Simulation results on gain (imaginary part of Ω) of PMI in normal dispersion regime [[© Ref. [14]].

[28]. However, if the fiber birefringence becomes sufficiently strong, the group velocity difference dominates, causing the polarization components to decouple and propagate independently. A representative spectrum of a GVL bright soliton formed under such conditions is shown in **Figure 5**.

The spectra of the GVL solitons have the following characteristics: (a) The spectrum of GVL solitons is typically broad due to the presence of XPC. (b) The central wavelength of the solitons along each polarization direction differs. The spectral properties vary depending on the type of GVL soliton formed. For example, in the case of bright GVL solitons and GVL dark solitons, two sets of Kelly sidebands can be observed due to that each of the coupled solitons has different central wavelengths. Besides, their spectral shapes are similar. However, for GVL dark-bright solitons, the spectral

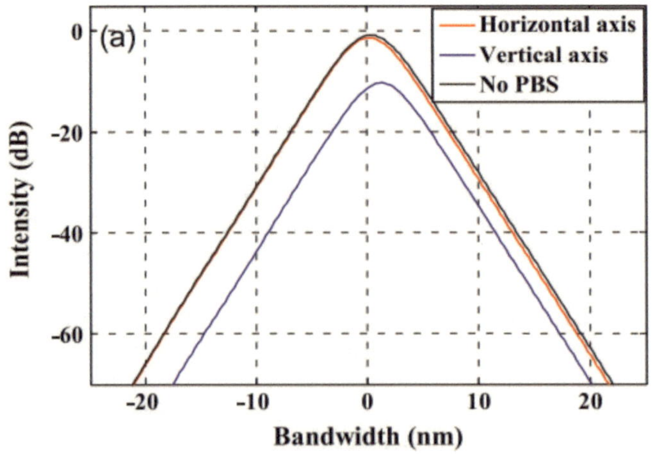

Figure 5.
Simulated spectra of the GVL bright solitons based on incoherent NLSE. [© Ref. [28]].

characteristics of the two solitons are distinctly different. Specifically, relating to the formed dispersion regime, the bright soliton could exhibit a broader spectral bandwidth than the dark soliton if the GVL dark-bright solitons are formed in the anomalous dispersion regime. This situation is reversed in the normal dispersion regime.

2.2.4 Coherent energy exchange FWM type of sidebands of vector solitons in SMFS

When the linear birefringence of a SMF is sufficiently small, the group velocity mismatch term in Eq. (17) could be neglected, and the effect of FWM becomes significant; thus, the pulse propagation is governed by the coherently coupled NLSEs. It is well-known that through the SPM and XPC between the two polarization modes of a fiber, the GVL solitons can be formed, as we discussed in Section 2.1.3. In addition to the formation of GVL solitons, it has been theoretically shown that the FWM coupling, also known as coherent energy exchange, between the polarization components of a vector soliton could have contributed to the formation of phase-locked vector solitons [29]. Further, different research groups have shown that the FWM coupling could induce new spectral sidebands, for instance, dip-peak type of spectral sidebands [6, 30]. A typical such formed spectra of the FWM sidebands is presented in **Figure 6**.

Numerical simulations have shown that weak linear birefringence within the cavity can give rise to four-wave mixing (FWM) between the two components of a vector soliton. This FWM interaction leads to energy transfer between the orthogonally polarized soliton components, resulting in the emergence of dip-peak-shaped spectral sidebands in the polarization-resolved soliton spectra. While coherent coupling enables such energy exchange, a nonzero birefringence restricts the interaction to specific wavelengths where phase-matching conditions are met in the cavity. This selective coupling gives rise to distinct additional spectral sidebands. Since FWM is a parametric process occurring internally between the soliton components, it induces only antiphase periodic intensity oscillations between the polarization modes without altering the overall intensity of the vector soliton itself.

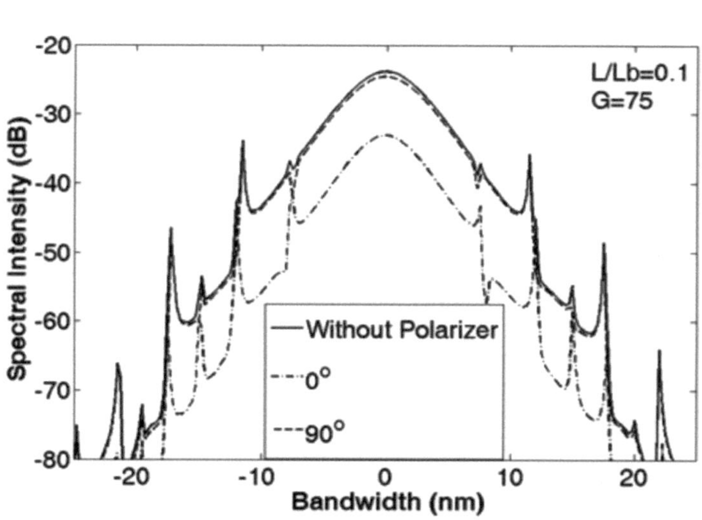

Figure 6.
Numerically simulated spectra of the phase-locked vector solitons in SMFs [© Ref. [30]].

3. Outlook on spectral sidebands instabilities in SMF lasers

Investigating sideband structures formed in fiber lasers and soliton dynamics is crucial for optimizing ultrafast laser performance. Key factors influencing these spectral features include soliton dynamics, higher-order dispersion-induced instabilities, and cross-phase modulation (XPM) in vector solitons.

Dip-type sidebands arise from a four-wave mixing (FWM) parametric interaction between solitons and dispersive waves, unlike traditional Kelly sidebands. Their spectral position varies significantly with linear cavity phase detuning, a dependency absent in conventional Kelly sidebands. Effectively controlling cavity phase detuning to enable precise spectral shaping of fiber lasers remains an open challenge and a topic of ongoing debate. Higher-order dispersion plays a critical role in sideband instability. While Kelly sidebands are primarily influenced by 2-nd order dispersion, third- and fourth-order dispersions alter phase-matching conditions, leading to spectral asymmetry, broadening, or additional sidebands. Understanding these effects is essential for mitigating pulse degradation in ultrafast lasers. Moreover, due to the vector nature of light, optical pulse propagation in a linear birefringent SMF is governed by the coupled NLSEs that admit different types of vector soliton solutions. In vector solitons, XPM couples polarization components, shifting sideband positions based on relative phase, polarization state, and power. This affects dispersive wave generation and requires detailed theoretical and experimental study for designing polarization-sensitive ultrafast lasers.

Future research should focus on numerical modeling and experimental techniques to better understand and control these phenomena. Advances in sideband engineering will improve fiber laser design for ultrafast optics, frequency comb generation, and high-speed optical communication.

4. Conclusions

In this chapter, we have theoretically investigated various types of spectral sidebands formed in single-mode soliton fiber lasers based on the nonlinear Schrödinger equation (NLSE) and the coupled NLSEs. Broadly, these spectral sidebands can be categorized into two main groups: linear and parametric sidebands. An example of linear sidebands is Kelly sidebands, which arises due to a linear interference effect. Their formation can be interpreted as the result of constructive interference between dispersive waves emitted from soliton perturbations as they propagate within the ring fiber cavity. The formation of Kelly sidebands is not a threshold effect. As far as a soliton is formed in fiber lasers, the Kelly sidebands are observed.

Unlike Kelly sidebands, which form as a thresholdless effect, the second group of sidebands, parametric sidebands, appears only when the light intensity exceeds a certain threshold. Furthermore, depending on the specific four-wave mixing (FWM) mechanisms, light propagation in single-mode fibers (SMFs) exhibits different behaviors, leading to the formation of modulation instability (MI)-induced spectral sidebands and/or dip-peak type soliton sidebands.

First, MI-induced sidebands can be regarded as an FWM-driven parametric energy transfer process between continuous-wave (CW) light and weak perturbations in anomalous dispersion regimes of optical fibers. When cavity-assisted effects are taken into account, this parametric energy transfer process can also extend to the normal dispersion regime, commonly referred to as the cavity-induced modulation instability (CIMI) effect.

Second, unlike MI-induced sidebands that are formed through a FWM process between CW light and weak modulations and whose onset does not require soliton formation, the FWM can also occur between solitons and dispersive waves when solitons are present inside a ring fiber cavity. This interaction leads to the emergence of dip-peak type sidebands. Notably, the location of dip-peak sidebands varies significantly with linear cavity detuning, a dependence that does not exist for Kelly sidebands. Moreover, if the formation of dip-peak sidebands is associated with soliton period-doubling or period-tripling bifurcations, the alternation of these sidebands must follow the same pulsation periods. For example, in the case of period doubling, a spectral dip transforms into a spectral peak after one round trip, whereas in the period tripling case, this transformation occurs after two round trips.

Finally, due to the vector nature of light, optical pulse propagation in a linearly birefringent single-mode fiber (SMF) is governed by the coupled nonlinear Schrödinger equations (CNLSEs). Compared to the scalar nonlinear Schrödinger equation (NLSE), the CNLSE introduces more complex nonlinear dynamics due to the coupling effects between orthogonally polarized modes. For instance, under incoherent coupling, modulation instability (MI) can also occur in the normal dispersion regime of the fiber. Meanwhile, under coherent coupling, polarization modulation instability (PMI) can arise in the same dispersion regime. Notably, unlike conventional MI or MI occurring under incoherent cross-phase coupling (XPC)—where MI gain peaks are always located at high modulation frequencies—under coherent cross-polarization coupling, PMI gain peaks can appear at low or even vanishing modulation frequencies. Moreover, if vector solitons form within the fiber cavity, coherent coupling can also induce dip-peak sidebands between the two polarization modes *via* the four-wave mixing (FWM) process. It is important to note that, since FWM is a parametric process occurring between the internal components of a vector soliton, it results only in an antiphase periodic intensity variation between the coupled soliton components, without affecting the overall intensity of the vector soliton [15]. For example, when measuring the total laser emission, no distinct spectral dips or peaks can be observed.

It is worth emphasizing that, in most cases, various nonlinear phenomenon coexist in fiber lasers. Therefore, a detailed analysis of the spectral sideband characteristics under different effects helps us effectively distinguish and identify the formation mechanisms of different sidebands. The research will contribute to better control over the formation of various sidebands, which in turn will be beneficial for the development of ultrafast lasers.

Acknowledgements

I would like to express my deepest gratitude to Prof. Yoo Seongwoo for his invaluable guidance, insightful feedback, and unwavering support throughout the process of writing this book chapter. Their expertise, patience, and encouragement were instrumental in shaping the content and refining the ideas presented here.

Author details

Hu Xiao[1,2] and Dingyuan Tang[1]*

1 Future Technology School, Shenzhen Technology University, China

2 School of Electrical and Electronic Engineering, Nanyang Technological University, Singapore

*Address all correspondence to: tangdingyuan@sztu.edu.cn

References

[1] Sulem PL. The Nonlinear Schrödinger Equations. Springer-Verlag; 1999. DOI: 10.1007/b98958

[2] Menyuk CR. Nonlinear pulse propagation in birefringent optical fibers. IEEE Journal of Quantum Electronics. 1987;**23**:174-176. DOI: 10.1109/JQE.1987.1073308

[3] Agrawal GP. Nonlinear Fiber Optics. Academic; 2012. DOI: 978-0-12-397023-7

[4] Kivshar YS, Agrawal GP. Optical Solitons: From Fibers to Photonic Crystals. Academic Press; 2003

[5] Copie F, Randoux S, Suret P. The physics of the one-dimensional nonlinear Schrödinger equation in fiber optics: Rogue waves, modulation instability and self-focusing phenomena. Reviews in Physics. 2020;**5**:100037

[6] Du Y, Shu X, Cao H, Cheng P. Dynamics of dispersive wave and regimes of different kinds of sideband generation in mode-locked soliton fiber lasers. IEEE Journal of Selected Topics in Quantum Electronics. 2018;**24**:1-8. DOI: 10.1109/JSTQE.2017.2777531

[7] Agrawal GP. Modulation instability induced by cross-phase modulation. Physical Review Letters. 1987;**59**: 880-883. DOI: 10.1103/PhysRevLett.59.880

[8] Tai K, Hasegawa A, Tomita A. Observation of modulational instability in optical fibers. Physical Review Letters. 1986;**56**:135-138. DOI: 10.1103/PhysRevLett.56.135

[9] Coen S, Haelterman M. Modulational instability induced by cavity boundary conditions in a normally dispersive optical fiber. Physical Review Letters. 1997;**79**:4139-4142. DOI: 10.1103/PhysRevLett.79.4139

[10] Tang DY, Guo J, Song YF, Li L, Zhao LM, Shen DY. GHz pulse train generation in fiber lasers by cavity induced modulation instability. Optical Fiber Technology. 2014;**20**:610-614. DOI: 10.1016/j.yofte.2014.08.002

[11] Dennis ML, Duling IR III. Experimental study of sideband generation in femtosecond fiber laser. IEEE Journal of Quantum Electronics. 1994;**30**:6. DOI: 10.1109/3.299472

[12] Wu QC, Liu C, Yao Y, Wu CH, Xu XC, Yang YF, et al. Experimental and simulated demonstration of a theoretical model for dip-type spectral sidebands in soliton fiber lasers. Optics & Laser Technology. 2021;**141**:107152. DOI: 10.1016/j.optlastec.2021.107152

[13] Wabnitz S. Modulational polarization instability of light in a nonlinear birefringent dispersive medium. Physical Review A. 1988;**38**: 2018-2021. DOI: 10.1103/PhysRevA.38.2018

[14] Chiu HS, Chow KW. Effect of birefringence on the modulation instabilities of a system of coherently coupled nonlinear Schrödinger equations. Physical Review A. 2009;**79**: 065803. DOI: 10.1103/PhysRevA.79.065803

[15] Liu M, Chen H, Luo A, Zhou G, Luo Z, Xu W. Identification of coherent and incoherent spectral sidebands in an ultrafast fiber laser. IEEE Journal of Selected Topics in Quantum Electronics. 2018;**24**:1-6. DOI: 10.1109/JSTQE.2017.2727881

[16] Thompson JR, Roy R. Statistical fluctuations in multiple four-wave mixing in a single-mode optical fiber. Physical Review A. 1991;**44**:7605. DOI: 10.1103/PhysRevA.44.7605

[17] Hasegawa A, Brinkman W. Tunable coherent IR and FIR sources utilizing modulational instability. IEEE Journal of Quantum Electronics. 1980;**16**:7. DOI: 10.1109/JQE.1980.1070554

[18] Tai K, Hasegawa A, Tomita A. Observation of modulational instability in optical fibers. Physical Review Letters. 1986;**56**:2. DOI: 10.1103/PhysRevLett.56.135

[19] Matera F et al. Sideband instability induced by periodic power variation in long-distance fiber links. Optics Letters. 1993;**18**:1499-1501. DOI: 10.1364/OL.18.001499

[20] Haelterman M, Trillo S, Wabnitz S. Dissipative modulation instability in a nonlinear dispersive ring cavity. Optics Communication. 1992;**91**:401-407. DOI: 10.1016/0030-4018(92)90367-Z

[21] Haelterman M, Trillo S, Wabnitz S. Additive-modulation-instability ring laser in the normal dispersion regime of a fiber. Optics Letters. 1992;**17**:745-747. DOI: 10.1364/OL.17.000745

[22] Tang et al. Cavity-assisted modulation instability lasing of a fiber ring laser. Applied Physics B. 2019;**125**:5. DOI: 10.1007/s00340-018-7117-4

[23] Kelly SMJ. Characteristics sideband instability of periodically amplified average soliton. Electronics Letters. 1992;**28**:455-457. DOI: 10.1049/el:19920508

[24] Agrawal GP. Modulation instability induced by cross-phase modulation. Physical Review Letters. 1987;**59**:880-883

[25] Yoshida E, Nakazawa M. Low-threshold 115-GHz continuous-wave modulational-instability erbium-doped fiber laser. Optics Letters. 1997;**22**:1409-1411. DOI: 10.1364/OL.22.001409

[26] De Matos CJS, Chestnut DA, Taylor JR. Low-threshold self-induced modulational instability ring laser in highly nonlinear fiber yielding a continuous-wave 262-GHz soliton train. Optics Letters. 2002;**27**:915-917. DOI: 10.1364/OL.27.000915

[27] Menyuk CR. Stability of solitons in birefringent optical fibers. I: Equal propagation amplitudes. Optics Letters. 1987;**12**:614-616. DOI: 10.1364/OL.12.000614

[28] Cao XD, McKinstrie CJ. Solitary-wave stability in birefringent optical fibers. Journal of the Optical Society of America B: Optical Physics. 1993;**10**:1202-1207. DOI: 10.1364/JOSAB.10.001202

[29] Collings BC et al. Polarization-locked temporal vector solitons in a fiber laser: Theory. Journal of the Optical Society of America B: Optical Physics. 2000;**17**:366-372. DOI: 10.1364/JOSAB.17.000354

[30] Zhang H, Tang DY, Zhao LM, Xiang N. Coherent energy exchange between components of a vector soliton in fiber lasers. Optics Express. 2008;**17**:12618-12623. DOI: 10.1364/OE.16.012618

Chapter 7

Mid-Infrared Laser Sources and Their Applications in Materials Processing

Kenji Goya, Seiji Shimizu, Ryo Yasuhara, Shigeki Tokita and Hiyori Uehara

Abstract

This review focuses on recent advances in mid-infrared (MIR) laser technology in materials processing, with an emphasis on fiber Bragg gratings (FBGs) fabricated by femtosecond lasers. The plane-by-plane femtosecond writing technique enables precise, highly stable FBGs directly within fluoride glass fibers, forming compact, wavelength-selective laser cavities. These advances have led to the development of high-efficiency MIR fiber lasers operating near 2.8–2.92 μm, offering strong material absorption that facilitates cutting, surface modification, and fusion splicing with minimal thermal damage. Polymer and glass materials, which often pose challenges for traditional near-infrared (NIR) lasers, can be processed more effectively with MIR sources. Although FBGs are integral to many MIR laser architectures, certain systems such as 2.92 μm Er:YAP lasers have been realized without FBGs. MIR fiber lasers present a promising platform for photonic manufacturing, addressing the growing demand for precise and efficient processing across a variety of advanced materials. Continued innovations in FBG fabrication, fiber integration techniques, and thermal management are expected to further expand the capabilities and applications of MIR laser technologies.

Keywords: femtosecond laser, FBGs, fluoride fiber, fiber laser, mid-IR, plastic fiber

1. Introduction

Mid-infrared (MIR) lasers have significantly advanced the landscape of material processing, particularly with the emergence of pulsed laser technologies. Many materials, including polymers and glasses, exhibit strong linear absorption in the MIR region—particularly around 2.8–3.0 μm—due to molecular vibration modes such as OH and CH stretching. This fundamental property has opened new pathways for precision thermal processing, in contrast to traditional ultraviolet (UV) and near-infrared (NIR) lasers that often require high fluences or the use of absorbing additives.

Recent progress in MIR laser development [1–15], including rare-earth-doped solid-state lasers and fiber-based architectures, has led to the realization of continuous-wave (CW) output powers exceeding 40 W [10]. Combined with improvements in beam quality and wavelength tunability [16–19], these systems are

IntechOpen

increasingly positioned as viable tools for advanced manufacturing. Notably, unlike 1 μm class lasers such as Nd:YAG—whose limited absorption in dielectric materials often necessitates the use of absorptive dopants [20]—MIR lasers naturally couple energy into transparent materials like polymers and glasses without modification, enabling efficient cutting, fusion bonding [21], and surface treatment [22] with minimal thermal damage.

A key enabling technology for MIR fiber lasers is the fiber Bragg grating (FBG) [11, 19, 23–28], which acts as an in-fiber mirror for wavelength selection and mode control. The femtosecond laser inscription of FBGs *via* the plane-by-plane method has proven especially powerful, allowing precise control over the grating pitch, index modulation, and length, even through polymer buffer coating. We demonstrated *in situ* inscription of first-order FBGs in Er-doped fluoride fibers for lasing at 2.8 μm, monitoring the real-time evolution of Bragg reflection and laser output during grating formation [26, 29]. Their work achieved 97% reflectivity in a 2.5-mm grating, enabling highly efficient lasing near the atmospheric transmission window (2799–2800 nm), with slope efficiencies over 29%.

In parallel with these laser oscillator developments, MIR laser systems are finding increasing use in materials processing applications, especially where NIR lasers are limited in their applicability due to the low absorption of certain materials at near-infrared wavelengths. For example, a CW fiber laser operating at 2.8 μm was employed to realize adhesive-free fusion bonding of PMMA-based plastic optical fibers (POFs) [21]. By adjusting laser power, spot size, and alignment, we achieved mechanically robust splices with acceptable optical transmission characteristics. Comparative analysis between 1 and 2.8 μm wavelengths revealed that MIR light enables over 20× improvement in thermal coupling efficiency, highlighting the wavelength's suitability for polymer joining.

Furthermore, MIR lasers were explored for their applicability in glass processing. In the case of alkali-free glass, finite element simulations suggested that 2.8 μm laser irradiation could enable efficient cutting with low thermal loading, outperforming CO_2 lasers (10.6 μm), which typically overheat the material due to high absorption and shallow penetration depths. In contrast, experimental validation was conducted for surface finishing using mid-IR laser chamfering, which successfully removed micro-chipping and residual stress, enhancing mechanical integrity as evidenced by improved bending strength [30, 31].

Additionally, we have applied a CW Er:YAP laser (2.92 μm) [32] to study the cutting of polymer films with varied absorption coefficients. The results reveal a clear dependence of cutting speed on optical absorption and thermal properties [33, 34]. Highly absorptive materials such as PET and PMMA exhibited efficient processing with well-defined cut profiles, while low-absorbing polymers such as PC and PTFE required lower scan speeds or failed to cut, due to poor heat accumulation and energy diffusion.

Taken together, these results demonstrate that MIR lasers not only advance fiber laser technologies through innovative FBG fabrication but also enable high-precision, application-specific material processing across polymers and glasses. This chapter surveys the progress of FBG-based MIR laser development, then shifts to practical applications in polymer cutting, fiber splicing, and glass treatment, illustrating the broad utility of MIR laser systems in modern photonic engineering.

The following sections provide an overview of FBG-based MIR laser development and present selected applications in polymer and glass processing, highlighting the advantages of MIR wavelengths for thermal control and material selectivity.

2. Fabrication of fiber Bragg gratings for mid-infrared fiber lasers

The fiber Bragg grating (FBG) is an essential component in the development of narrow-linewidth, wavelength-stabilized mid-infrared (MIR) fiber lasers. FBGs inscribed in rare-earth-doped fluoride glass fibers, such as Er:ZBLAN, enable cavity feedback directly within the fiber medium, allowing for monolithic laser architectures with enhanced robustness and compactness.

Among various fabrication techniques, the plane-by-plane femtosecond laser inscription method has become the most versatile and widely adopted approach. This method allows precise control over the grating pitch, refractive index modulation, and grating length. As shown in **Figure 1**, it is also capable of writing through polymer coatings without damaging the fiber, a key advantage for practical deployment.

Our work demonstrated real-time monitoring of the lasing evolution during FBG inscription in Er-doped fluoride fibers operating at 2.8 μm [26]. By observing the transition from amplified spontaneous emission (ASE) to stable laser oscillation, the grating parameters was optimized *in situ*, achieving high reflectivity (>97%) in gratings as short as 2.5 mm, as can be seen in **Figure 2a**. **Figure 2b** shows the FBG reflectivity R_B simply modeled using the following equation:

$$R_B = \tanh\left(\frac{\pi \delta_n L}{\lambda}\right) + TR_F \tag{1}$$

where δ_n is the refractive index change, R_F is the Fresnel reflectivity, and T is the FBG transmittance. Light reflected by the Fresnel reflection at the fiber end (right side of **Figure 1**) is expressed as TR_F. The calculation is in good agreement with the experimental result when the calculation is performed assuming $\delta_n = 1.1 \times 10^{-3}$ at $\lambda = 2799.7$ nm. The index change is comparable to the value reported in [23].

Furthermore, the selection of the Bragg wavelength in the MIR region—particularly near 2.8 μm—can be tuned to match the absorption peaks of target materials. This wavelength selectivity is crucial for enabling efficient energy coupling into

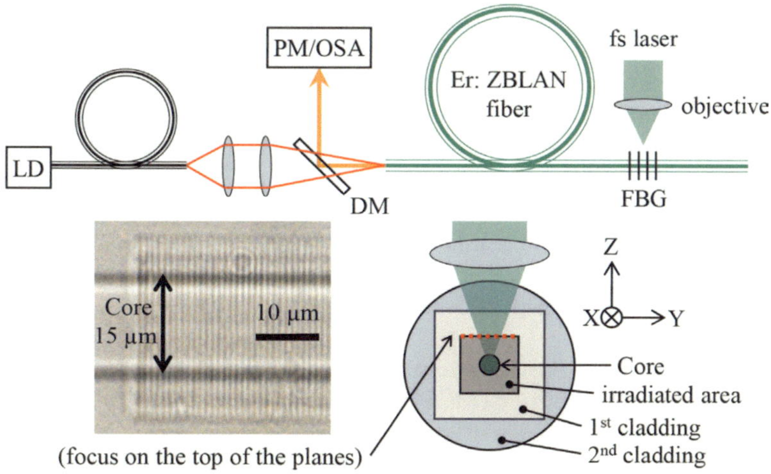

Figure 1.
Experimental setup for FBG inscription and laser monitoring. LD, laser diode; PM, power meter; OSA, optical spectrum analyzer, DM, dichroic mirror. (Reproduced from our previous work [26]).

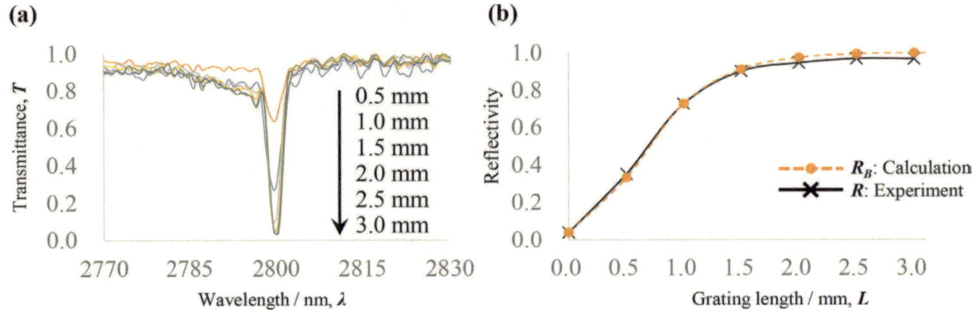

Figure 2.
(a) Transmission spectra measured for various grating lengths. Resolution: 350 pm. (b) FBG peak reflectivity at each grating length. (Adapted from our previous work [26]).

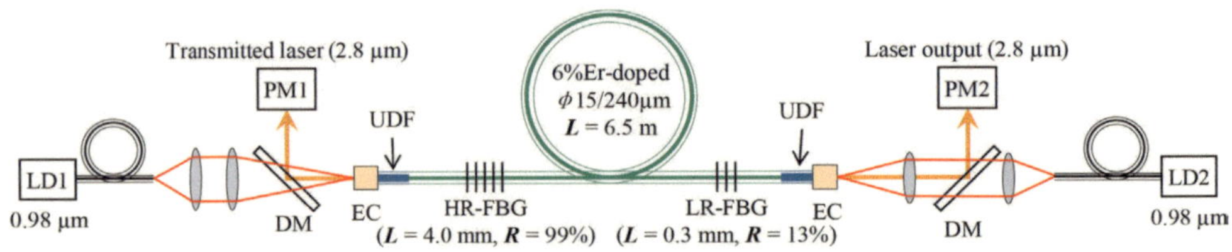

Figure 3.
Schematic drawing of 35 W fiber laser. EC, endcap; UDF, undoped fiber; DM, dichroic mirror; L1, L2, L3 and L4, coupling lenses; PM, power meter. (Reproduced from our previous work [4]).

polymers and glasses during laser processing applications, which is explored in the subsequent chapters.

Further demonstrating the maturity of this technology, we reported the realization of a high-power Er:ZBLAN fiber laser (see **Figure 3**) [4] operating at 2838 nm with a maximum output of 35 W and continuous 30 W operation for over an hour. In this system, fiber Bragg gratings were directly inscribed into the core of the Er-doped fluoride fiber using the plane-by-plane femtosecond technique—without removing the polymer coating—achieving excellent thermal durability. The development of key components, including FBGs, fusion splicing methods for fluoride fibers, and CaF_2 endcaps, was central to ensuring long-term output stability and high beam quality. Notably, the FBGs served not only as cavity reflectors but also as diagnostic points, allowing *in situ* monitoring of transmitted power during laser operation to assess grating performance and detect potential degradation. These achievements underscore the critical role of robust FBG fabrication and integration techniques in enabling scalable, stable, and efficient mid-infrared fiber laser systems. The combination of high reflectivity, compact cavity design, and thermal management strategies positions FBG-based MIR fiber lasers as promising platforms for both scientific and industrial applications.

In addition to oscillator-based systems, a power-scalable architecture known as the master oscillator power amplifier (MOPA) [2] has gained attention for high-power mid-infrared applications. This configuration combines a stable seed oscillator—often incorporating FBG-based cavity feedback—with a high-gain fiber amplifier stage. The MOPA structure is particularly effective in overcoming the thermal and mechanical limitations of fluoride fibers, enabling greater power scalability while preserving beam quality and wavelength stability. As can be seen from **Figure 4**, a representative

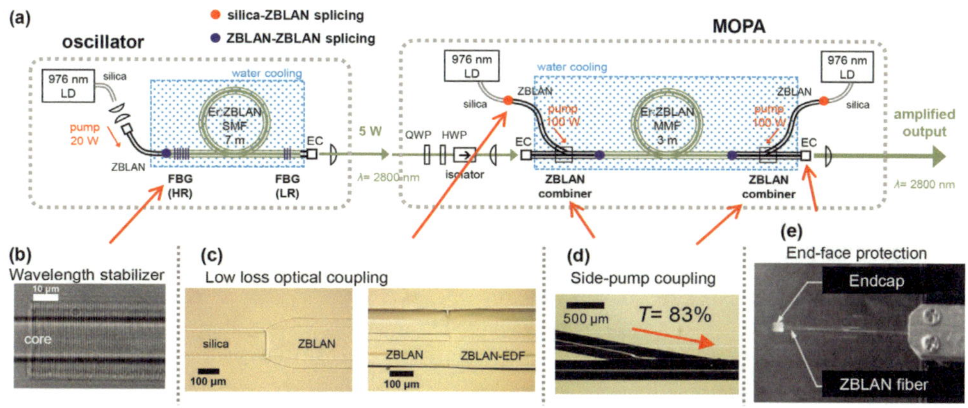

Figure 4.
(a) Schematic diagram of the setup for an Er:ZBLAN fiber MOPA, Optical microscope images of (b) FBGs (c) fusion-spliced fibers and (d) endcaps. (Adapted from our previous work [2]).

example is the Er:ZBLAN fiber MOPA system reported by Uehara et al., which achieved 33.1 W of amplified output at 2.8 μm with slope efficiencies of up to 28%. The system employs high-quality FBGs inscribed *via* plane-by-plane femtosecond laser processing for stable wavelength control. Moreover, it incorporates advanced fusion splicing techniques between silica and fluoride fibers to reduce coupling loss, and side-pump combiners to enable efficient bidirectional pumping [35]. These pump combiners, fabricated *via* thermal fusion, allow spatially distributed excitation over long gain fibers, thereby mitigating local thermal loading.

Such technological advancements—including FBG design, robust fiber interfacing, and innovative pump delivery methods—are critical for building high-power MIR fiber systems with long-term stability. The MOPA architecture provides a flexible and modular foundation for future industrial grade laser platforms and forms a bridge between scientific research and practical applications in cutting, welding, and sensing.

3. Absorption characteristics of resin materials

In recent years, multilayered resin films with high added value have been widely adopted in display panels, such as those used in smartphones. These materials demand high-precision processing technologies. Mid-infrared (MIR) lasers, particularly in the 2.8–3.0 μm range, exhibit strong absorption in many transparent polymer materials due to molecular vibrational modes such as OH and CH stretching. This makes them promising candidates for high-throughput cutting and processing applications. Although the mechanisms of laser interaction differ between polymers and glass, the fundamental process is similar: light is absorbed by the material, leading to localized heating, thermal expansion or melting, and eventually resulting in shape transformation. Therefore, the optimization of laser absorption properties is critical for efficient and precise material processing.

Figure 5 shows the infrared transmittance spectra of typical resin materials (thickness: 0.1–0.2 mm) used in display panels and related applications. A lower transmittance indicates higher material absorption. For instance, polyethylene terephthalate (PET) exhibits strong absorption peaks around 2.92 μm—the emission wavelength of the Er:YAP laser—primarily attributed to the stretching vibrations of

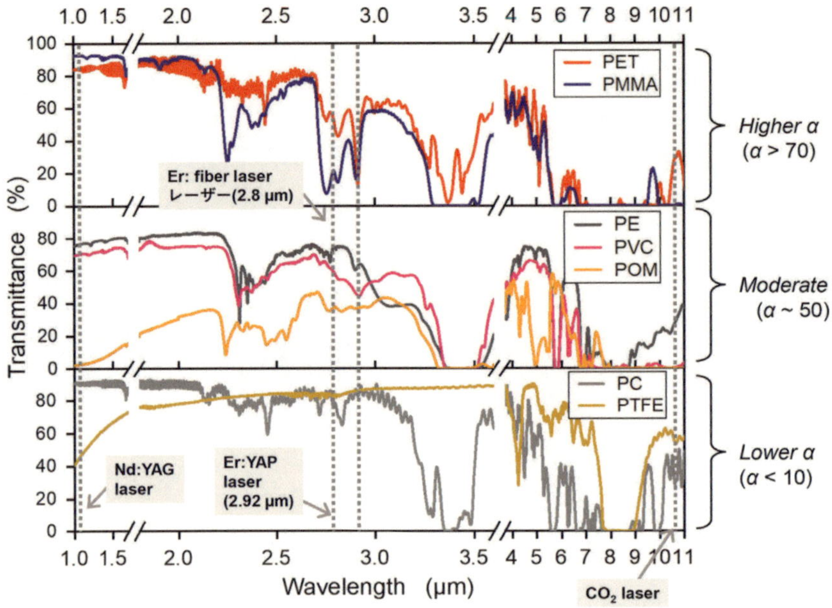

Figure 5.
Transmittance spectra of typical resin materials.

Sample designation	$\alpha_{\lambda = 2.92\ \mu m}$ [cm^{-1}]	T_g, [°]	γ, [mN/m]	a_l [10^{-5}/°C]	k [W/m K]
PET	160	250	43	2	0.15
PMMA	70	100	42	5–9	0.17–0.25
PE	40	100–140	31	11–18	0.34–0.50
PVC	50	100–200	39	5–25	0.13–0.29
POM	50	180	42	8.1	0.23
PC	10	150	34	6.6	0.19
PTFE	10	330	18	2	0.25

Table 1.
Absorption coefficient and thermal property of resin materials.

OH and CH bonds. The absorption coefficient α for PET is estimated to be around 160 cm^{-1}, indicating strong absorption.

Table 1 summarizes the absorption coefficients of other resin materials at 2.92 μm, including polyethylene (PE), polymethyl methacrylate (PMMA), polyvinyl chloride (PVC), polyacetal (POM), polycarbonate (PC), and polytetrafluoroethylene (PTFE). For reference, the table also includes key thermal properties such as melting point T_g, surface tension γ, thermal expansion coefficient a_l, and thermal conductivity k. These parameters are essential for understanding laser-material interaction and the resulting thermal response.

By selecting materials with appropriate absorption characteristics and thermal behavior, mid-infrared laser systems can be effectively tailored for various resin processing tasks, as described in the subsequent sections.

4. Application to polymer processing

4.1 Fusion splicing of PMMA plastic optical fibers

This section describes a contactless fusion splicing technique for plastic optical fibers (POFs) made of polymethyl methacrylate (PMMA, $\alpha_{\lambda=2.8\mu m} = 82\ \text{cm}^{-1}$), with a core diameter of 0.5 mm. Unlike conventional methods that require adhesives or mechanical connectors, this approach utilizes mid-infrared laser radiation to thermally fuse fiber ends, offering a simplified and contamination-free process.

The fusion splicing system employs a custom-developed 2.8 μm continuous-wave (CW) fiber laser (maximum output: 5 W, beam quality: $M^2 < 1.0$) [21]. A calcium fluoride (CaF_2) lens with a focal length of 20 mm is used to focus the beam to a 15 μm spot size at the fiber end-face. Precise control over the irradiation time and spatial alignment with respect to the focal plane allows localized heating of the fiber ends for splicing.

Figure 6 illustrates the fusion procedure and the corresponding optical micrographs. First, the two fiber ends are preheated at 80 mW, brought into contact, and subsequently fused by increasing the laser power to 120 mW while translating one fiber end by a preset distance (Δx). During this process, fine adjustments to the alignment are made to maximize transmitted optical power, at which point the laser is turned off to complete the fusion. This method effectively minimizes splice loss between POF_1 and POF_2 while maintaining mechanical integrity.

In the experiment, the effects of Δx variation on optical transmission, allowable bending radius, and tensile strength were investigated. As shown in **Figure 7a**, transmission measurements were performed using a laser diode (LD) and photodiode (PD) setup, with five samples per condition. By comparing transmitted power before and

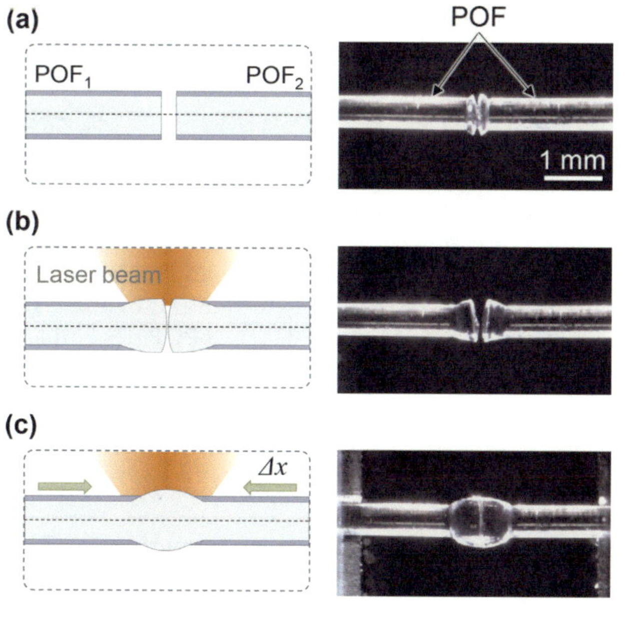

Figure 6.
Diagrams and photographs showing the steps of fusion splicing in this work. (a) Alignment of the POFs, (b) preheating to smooth the fiber ends, and (c) heating to a higher temperature while forcing the fibers together. (Adapted from our previous work [21]).

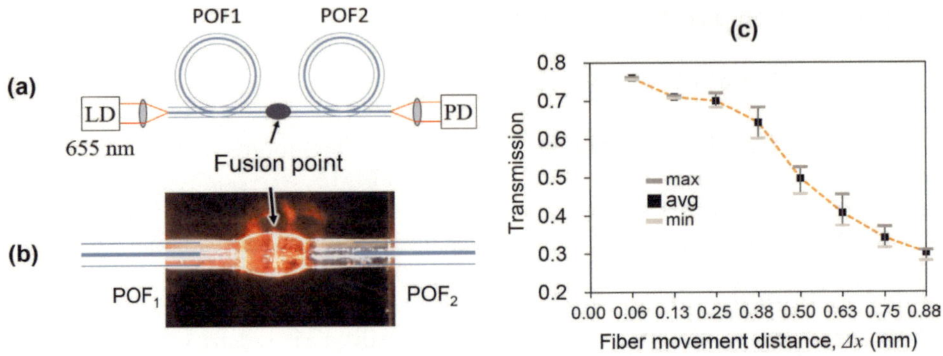

Figure 7.
(a) Experimental setup for evaluating the optical transmission of fusion-spliced plastic optical fibers (POFs). A 655 nm laser diode (LD) was used as the light source, and the transmitted signal was detected by a photodiode (PD). (b) Optical photograph of the fusion point between POF$_1$ and POF$_2$ during laser diode transmission. (c) Measured optical transmission as a function of fusion displacement Δx. (Adapted from our previous work [21]).

after fusion, intrinsic splice loss was isolated from coupling and propagation losses. **Figure 7b** presents a microscopic image of the fusion zone under LD illumination, where optical leakage was visualized. The transmission correlated with Δx, and as shown in **Figure 7c**, reducing Δx improved transmission by minimizing modal mismatch. The maximum transmission was 76% at Δx = 0.063 mm, the lowest manually controllable value in this setup. As Δx increased, the variation between maximum and minimum transmission narrowed, indicating improved reproducibility.

The allowable bending radius, defined as the minimum radius before fiber breakage, decreased with larger Δx, reaching below 1.5 mm for $\Delta x \geq 0.375$ mm. This behavior is attributed to the increase in joint diameter, which enhances resistance to mechanical stress. Tensile strength was evaluated using a force gauge with the fibers clamped at both ends. The measured strength increased with Δx, reaching a maximum of 13.5 N. Beyond this load, the fiber itself, not the splice, underwent plastic deformation, suggesting that 13.5 N represents the upper strength limit for the unmodified POF. When comparing cross-sectional areas, the tensile strength at the fusion point was calculated as 4.92 N/mm^2, while the original fiber exhibited 17.2 N/mm^2, demonstrating that the fusion bond provides sufficient mechanical strength for practical use.

In summary, this MIR-based fusion splicing technique successfully joins PMMA POFs without adhesives or complex processing. Although the resulting optical performance may not meet the requirements for high-end telecommunications, the achieved mechanical durability—particularly tensile and bending strength—is sufficient for practical implementation in short-distance POF networks. Furthermore, replacing manual Δx control with motorized stages is expected to improve optical transmission and process consistency.

4.2 Cutting of resin films

This section presents the results of laser cutting experiments [33, 34] conducted on various resin films using a custom-developed continuous-wave (CW) Er:YAP laser emitting at a wavelength of 2.92 μm [32]. The goal was to evaluate the feasibility and characteristics of mid-infrared laser processing across different polymer materials. It should be noted that while 2.8 μm fiber lasers incorporated FBGs for wavelength control, the 2.92 μm laser system employed in our experiments did not utilize FBG technology. Although a 2.92 μm Er:YAP laser was employed in this study, it is feasible to construct laser systems at similar wavelengths, such as 2.91 μm, using fiber-based architectures incorporating fiber Bragg gratings (FBGs) for wavelength selection and cavity formation [36].

The laser system used in the experiment delivered a maximum output of 2 W with a beam quality of $M^2 < 1.8$. The beam was focused through a plano-convex lens with a 40 mm focal length, achieving a spot diameter of approximately 30 μm on the sample surface. Each resin film was mounted on an automated translation stage and scanned linearly across the laser beam, as illustrated in **Figure 8a**.

The scanning speed was varied to identify the minimum energy density required for successful cutting. The results are summarized in **Table 2**, with cross-sectional profiles of the cut regions shown in **Figure 8b** and **c**. For highly absorptive materials such as polyethylene terephthalate (PET) and polymethyl methacrylate (PMMA), the maximum cutting speeds (denoted as v_{th}) were 2.0 and 0.5 cm/s, respectively. In contrast, for materials with moderate absorption coefficients, such as polyethylene (PE) and polyvinyl chloride (PVC), the cutting speed needed to be reduced to 0.1–0.2 cm/s to achieve continuous cuts, as shown in **Figure 9**.

In cases where the irradiation conditions were insufficient for cutting, swelling or deformation was observed at the beam center and surrounding areas. Particularly for materials with absorption coefficients below 10 cm^{-1}, such as polycarbonate (PC) and polytetrafluoroethylene (PTFE), no cutting was achieved under any tested condition.

To analyze the cutting behavior quantitatively, the irradiance per unit spot area was calculated using Eq. (2):

$$E = \frac{P}{v \times S} \tag{2}$$

where E is the energy density (W/cm^2), P is the laser power, v is the scan speed, and S is the spot area. **Figure 10** plots the energy density against absorption characteristics, showing a general correlation with cutting performance.

However, notable deviations were observed. For instance, although PVC and POM exhibit similar absorption coefficients, differences in the cross-sectional profiles and required energy densities were apparent. In the case of PVC, as shown in **Figure 9** (PVC), heat-affected zones were clearly visible around the cut edge. This suggests that

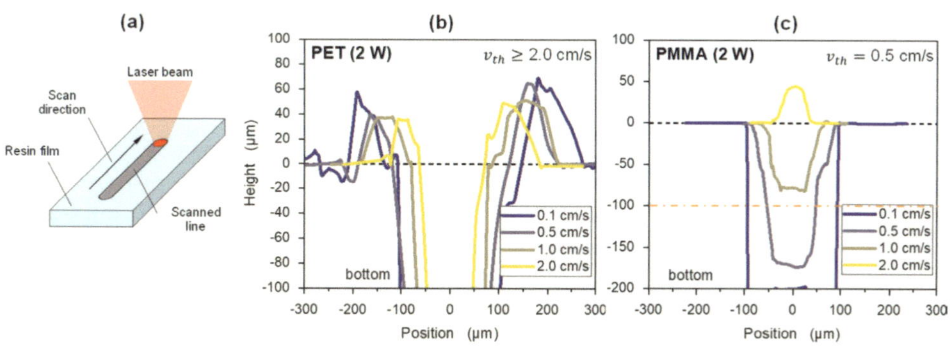

Figure 8.
(a) Schematic diagram of the cutting process. Cross-sectional profile of irradiated area on (b) PET and (c) PMMA. (Adapted from our previous work [33, 34]).

Sample designation	$\alpha_{\lambda = 2.92\ \mu m}$ [cm^{-1}]	t, [mm]	v_{th}, [cm/s]	P, [W]	E, [MW/cm^2]
PET	160	0.1	2	2	0.14
PMMA	70	0.2	0.5		0.57
PE	40	0.1	0.1		2.83
PVC	50	0.2	0.1		2.83
POM	50	0.2	0.2		1.41
PC	10	0.1	—		N/A
PTFE	10	0.2	—		N/A

Table 2.
Summary of experimental results.

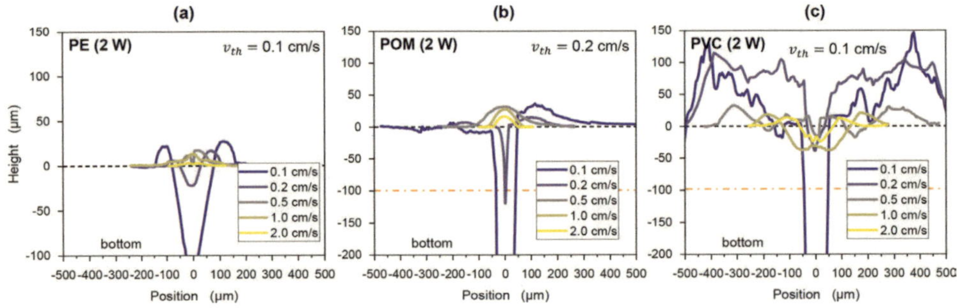

Figure 9.
Cross-sectional profiles of laser-irradiated areas on (a) PE, (b) POM, and (c) PVC. (Adapted from our previous work [33, 34]).

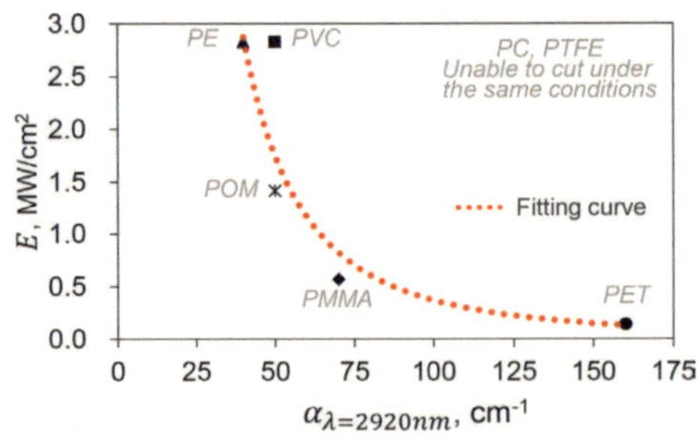

Figure 10.
Relationship between laser energy density per unit spot area (E, MW/cm^2) and absorption coefficient (cm^{-1}) for various polymer materials. Each marker represents a different material: PE (▲), PVC (■), POM (✱), PMMA (◆), and PET (●). The dotted red curve shows the fitting trend. PC and PTFE could not be cut under the same irradiation conditions.

during the thermal decomposition process, the laser energy was more prone to diffusing into surrounding areas rather than being confined to the focal region.

These findings indicate that the success of MIR-based polymer cutting is strongly influenced not only by absorption coefficients but also by thermal conductivity, melting point, and the material's tendency to undergo thermal deformation. Optimization of processing parameters—including wavelength selection, power density, and scan speed—along with auxiliary techniques such as assist gas flow, will be critical for achieving high-quality, high-throughput cutting performance across diverse polymer types.

5. Applications in glass processing

Mid-infrared (MIR) fiber lasers also show great promise in the field of glass material processing, where precise thermal control is essential. In this section, we introduce two major applications: cutting of alkali-free glass and surface finishing of diamond-cut edges, both utilizing a 2.8 μm CW fiber laser. Efficient laser cutting of glass requires an optimal balance between sufficient absorption and minimal thermal loading. Before discussing specific laser processing techniques, it is important to understand the thermal stress behavior of glass. As shown in **Figure 11**, owing to rapid cooling during/after laser irradiation, tensile stress develops at the surface while compressive stress accumulates inside the material. When the tensile stress exceeds the fracture threshold, cleavage will occur along the internal planes of weakness, such as microcracks or structurally weak regions that serve as natural fracture paths under tensile stress.

5.1 Cutting of alkali-free glass

As displayed in **Figure 12**, numerical simulations using the finite element method revealed that irradiation with a 2.8 μm laser results in a well-confined temperature distribution in alkali-free glass (thickness: 0.7 mm, spot size: φ = 2 mm), leading to minimal energy diffusion and effective cutting. This is primarily due to the moderate absorption coefficient of alkali-free glass at 2.8 μm (approximately 13 cm^{-1}), which enables controlled surface heating without causing excessive thermal stress or deep penetration. In contrast, CO_2 lasers operating at 10.6 μm have much higher absorption (approx. 360 cm^{-1}), which often leads to overly localized heating and insufficient depth control, especially in brittle materials like glass. Compared to infrared lasers, 1 μm lasers require a considerably higher energy input to achieve effective processing.

The 2.8 μm wavelength thus represents an ideal balance between sufficient absorption for efficient laser processing and limited thermal penetration that reduces the risk of cracking or thermal damage, enabling precise, clean cuts. These characteristics make MIR fiber lasers particularly suitable for applications where dimensional precision and surface quality are critical, such as in the production of glass substrates for optical or electronic components.

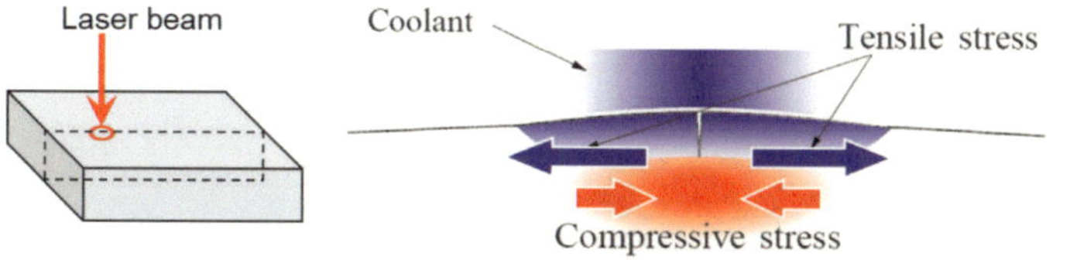

Figure 11.
Schematic illustration of stress distribution during cooling-induced cleavage during/after laser irradiation in glass. Rapid cooling from the surface generates tensile stress in the interior, leading to spontaneous fracture along internal planes.

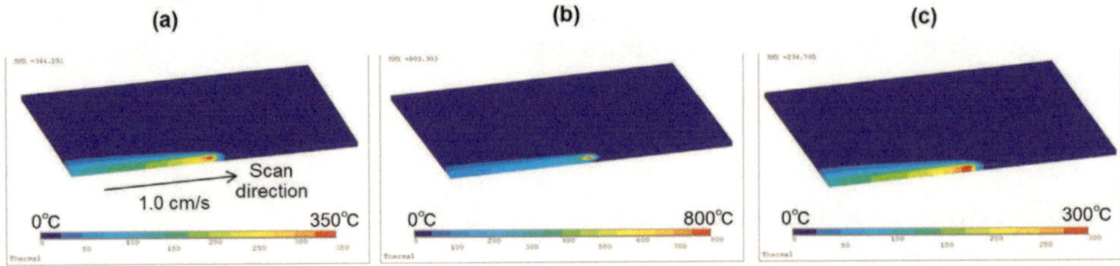

Figure 12.
Simulated temperature distributions in alkali-free glass during laser scanning at a speed of 1.0 cm/s under different laser conditions: (a) 2.8 μm laser at 10 W, (b) CO_2 laser at 5.7 W (not suitable for cutting), (c) 1.0 μm laser at 430 W. The 2.8 μm laser produces localized surface heating with a peak temperature around 350°C, while the CO_2 and 1.0 μm lasers result in deeper thermal penetration or insufficient surface heating for effective cutting.

5.2 Surface finishing by laser irradiation

In addition to cutting, MIR laser irradiation can be used for surface conditioning of mechanically processed glass. For instance, alkali-free glass samples cut using a diamond blade were irradiated from one direction with a 2.8 μm CW fiber laser (output: 4 W, scan speed: 0.3 cm/s), and the processed edges were subsequently observed from two directions, as shown in **Figure 13**.

The results demonstrated successful removal of micro-chipping and edge defects, resulting in a smoother, more uniform surface finish. Moreover, residual stress introduced during mechanical cutting was significantly mitigated. Four-point bending tests revealed that the B10 strength—defined as the load at which 10% of samples fail—was improved by nearly a factor of two after laser-based surface treatment. Chamfering quality was evaluated based on the ratio of chamfer width to glass thickness, as illustrated in **Figure 14**. The chamfer ratio R_C was defined as the total chamfer width divided by the glass thickness. Chamfering results were classified into four categories: excessive (△), good (○), small (◇), and insufficient (□). A good chamfer (○) corresponds to a chamfer ratio between 75 and 100%, indicating that the entire back surface of the glass is chamfered. In contrast, a chamfer ratio greater than 100% (△) indicates overprocessing, while ratios below 75% (◇ and □) reflect insufficient chamfering coverage.

Although the experiments presented here are preliminary, they underscore the potential of MIR fiber lasers to go beyond conventional mechanical or CO_2 laser-based

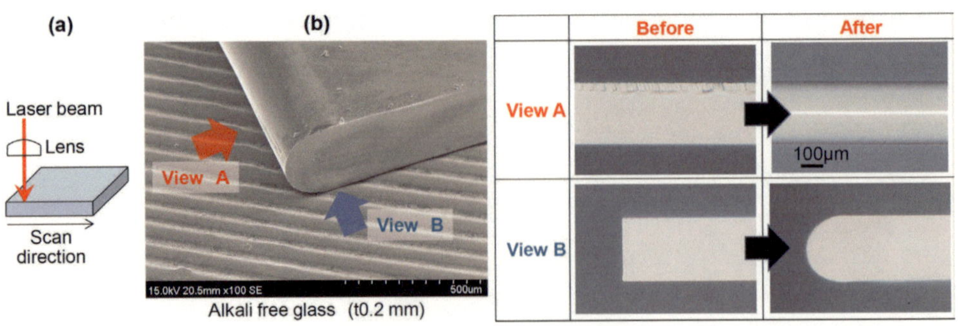

Figure 13.
Chamfering of a glass edge by laser melting. (a) Schematic of the laser irradiation process. (b) SEM image of the chamfered edge. Right panels show optical microscope images of the edge from two directions (View A and View B) before and after laser chamfering. (Adapted from our previous work [31]).

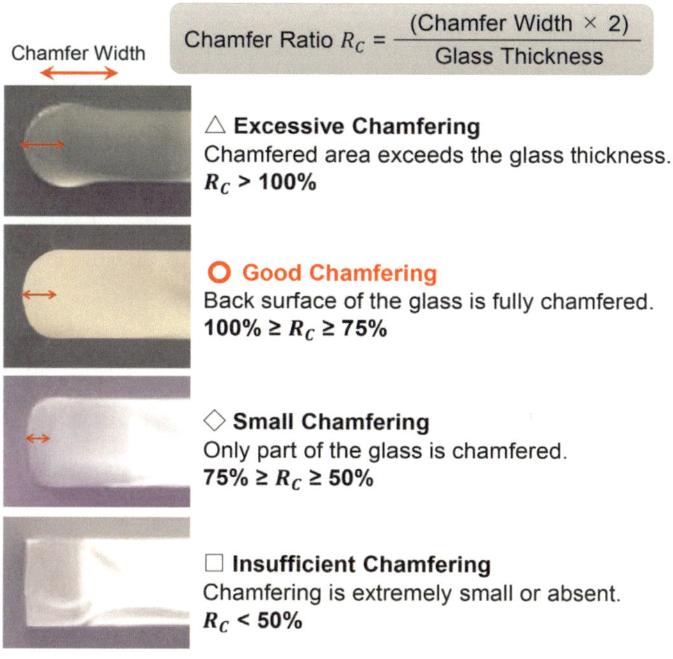

Figure 14.
C Classification of chamfering quality based on the chamfer ratio R_C. The classification is based on the degree of edge coverage achieved by laser processing. (Adapted from our previous work [30]).

approaches in glass processing. The ability to finely tune energy deposition and heat flow opens new pathways for defect reduction and strength enhancement, even in fragile materials.

6. Conclusion

This chapter has explored the potential of mid-infrared (MIR) lasers—particularly those operating in the 2.8–2.92 μm wavelength range—for advanced material processing applications, with a focus on the integration of FBGs in MIR fiber laser systems. Beginning with the fabrication of FBGs *via* the plane-by-plane femtosecond laser inscription method, we highlighted how compact, wavelength-selective, and efficient laser cavities can be realized within Er-doped fluoride glass fibers. These monolithic architectures offer high slope efficiencies, narrow linewidths, and stable output, forming the foundation for practical MIR laser sources.

Leveraging the strong vibrational absorption bands of many polymers and glasses in the MIR region, we demonstrated that FBG-assisted CW fiber lasers can serve as highly effective tools for thermal processing. In polymer applications, 2.8 and 2.92 μm lasers enabled adhesive-free fusion splicing of PMMA-based plastic optical fibers (POFs), achieving sufficient mechanical strength and acceptable optical transmission. Furthermore, laser cutting of resin films revealed a strong dependence on both absorption coefficient and thermal properties, with optimized parameters enabling clean, high-speed processing without excessive thermal damage.

In glass processing, the 2.8 μm wavelength provided an ideal balance between surface heating and penetration depth. This enabled efficient cutting of alkali-free glass substrates with minimal thermal loading and allowed laser-assisted chamfering to reduce micro-cracks and residual stress, thereby enhancing mechanical reliability.

Collectively, these results underscore the versatility of mid-infrared fiber lasers, not only as laser sources enabled by integrated FBGs but also as application-ready platforms for precision manufacturing. The ability to tailor emission wavelength, beam quality, and energy delivery—combined with insights into material-specific absorption behavior—makes MIR lasers uniquely positioned to address challenges in polymer and glass processing that are not well-served by traditional UV or NIR systems.

Looking ahead, further advances in FBG inscription techniques, automated fusion splicing controls, and thermal modeling will pave the way for widespread industrial adoption. As both the laser source and its processing applications mature, MIR fiber lasers are expected to play a key role in the next generation of photonic manufacturing technologies.

Acknowledgements

We acknowledge funding from JST A-STEP stage III type B (NexTEP B), Japan Science and Technology Agency "Adaptable and Seamless Technology Transfer Program through Target-driven R&D." This work was also partially supported by Kakenhi Grants-in-Aid from the Japan Society for the Promotion of Science (JSPS) and grants from Amada foundation.

Conflict of interest

This research was conducted in collaboration with Company Mituboshi Diamond Industrial Co., Ltd. The authors declare no competing interests related to the publication of this work.

Thanks

We gratefully acknowledge the contributions of former members of our project team, including Dr. Masanao Murakami, Dr. Satoshi Hattori, Mr. Daisuke Konishi, Dr. Christian A. Schäfer, Dr. Koichi Oda, Dr. Keisuke Yahata, Dr. Hiraku Matsukuma, Mr. Hideo Izumi, Dr. Ryo Sahara, Dr. Asuka Misawa, and Dr. Satoshi Tsuzuki.

Author details

Kenji Goya[1*], Seiji Shimizu[2], Ryo Yasuhara[3], Shigeki Tokita[4] and Hiyori Uehara[3]

1 Akita Prefectural University, Yurihonjo, Akita, Japan

2 Mituboshi Diamond Industrial Co., Ltd., Settsu, Osaka, Japan

3 National Institute for Fusion Science, Toki, Gifu, Japan

4 Institute for Chemical Research, Kyoto University Gokesho Ikenoura, Uji, Kyoto, Japan

*Address all correspondence to: k-gouya@akita-pu.ac.jp

References

[1] Tokita S, Murakami M, Shimizu S, Hashida M, Sakabe S. Liquid-cooled 24 W mid-infrared Er:ZBLAN fiber laser. Optics Letters. 2009;**34**:3062. DOI: 10.1364/ol.34.003062

[2] Uehara H, Konishi D, Goya K, Sahara R, Murakami M, Tokita S. Power scalable 30-W mid-infrared fluoride fiber amplifier. Optics Letters. 2019;**44**: 4777. DOI: 10.1364/ol.44.004777

[3] Tokita S, Hirokane M, Murakami M, Shimizu S, Hashida M, Sakabe S. Stable 10 W Er:ZBLAN fiber laser operating at 271–288 μm. Optics Letters. 2010;**35**: 3943. DOI: 10.1364/ol.35.003943

[4] Goya K, Uehara H, Konishi D, Sahara R, Murakami M, Tokita S. Stable 35-W Er: ZBLAN fiber laser with CaF_2 end caps. Applied Physics Express. 2019; **12**:102007. DOI: 10.7567/1882-0786/ab3f44

[5] Uehara H, Tokita S, Kawanaka J, Konishi D, Murakami M, Yasuhara R. A passively Q-switched compact Er:Lu_2O_3 ceramics laser at 2.8 μm with a graphene saturable absorber. Applied Physics Express. 2019;**12**:022002. DOI: 10.7567/1882-0786/aaf994

[6] Caron N, Bernier M, Faucher D, Vallée R. Understanding the fiber tip thermal runaway present in 3 μm fluoride glass fiber lasers. Optics Express. 2012;**20**:22188. DOI: 10.1364/oe.20.022188

[7] Bernier M, Faucher D, Caron N, Vallée R. Highly stable and efficient erbium-doped 28 μm all fiber laser. Optics Express. 2009;**17**:16941. DOI: 10.1364/oe.17.016941

[8] Maes F, Fortin V, Bernier M, Vallée R. 5.6 W monolithic fiber laser at 3.55 μm. Optics Letters. 2017;**42**:2054. DOI: 10.1364/ol.42.002054

[9] Fortin V, Bernier M, Bah ST, Vallée R. 30 W fluoride glass all-fiber laser at 2.94 μm. Optics Letters. 2015;**40**:2882. DOI: 10.1364/ol.40.002882

[10] Aydin YO, Fortin V, Vallée R, Bernier M. Towards power scaling of 2.8 μm fiber lasers. Optics Letters. 2018;**43**: 4542. DOI: 10.1364/ol.43.004542

[11] Faucher D, Bernier M, Androz G, Caron N, Vallée R. 20 W passively cooled single-mode all-fiber laser at 2.8 μm. Optics Letters. 2011;**36**:1104. DOI: 10.1364/ol.36.001104

[12] Fortin V, Jobin F, Larose M, Bernier M, Vallée R. 10-W-level monolithic dysprosium-doped fiber laser at 3.24 μm. Optics Letters. 2019;**44**: 491-494. DOI: 10.1364/OL.44.000491

[13] Aydin YO, Fortin V, Maes F, Valleé R, Bernier M. Long-term operation of high-power 3 μm fiber lasers. In: Advanced Solid State Lasers - Proceedings Laser Congress 2019 (ASSL, LAC, LS and C). Washington, D.C., USA: Optica Publishing Group; 2019. pp. 4-6. DOI: 10.1364/assl.2019.aw4a.7

[14] Paradis P, Fortin V, Aydin YO, Vallée R, Bernier M. 10 W-level gain-switched all-fiber laser at 2.8 μm. Optics Letters. 2018;**43**:3196. DOI: 10.1364/ol.43.003196

[15] Aydin YO, Maes F, Fortin V, Bah ST, Vallée R, Bernier M. Endcapping of high-power 3 μm fiber lasers. Optics Express. 2019;**27**:20659. DOI: 10.1364/oe.27.020659

[16] Majewski MR, Jackson SD. Tunable dysprosium laser. Optics Letters. 2016; **41**:4496. DOI: 10.1364/ol.41.004496

[17] Zhu X, Jain R. Compact 2 W wavelength-tunable Er:ZBLAN mid-infrared fiber laser. Optics Letters. 2007; **32**:2381. DOI: 10.1364/ol.32.002381

[18] Wang J, Zhu X, Norwood RA, Peyghambarian N. Widely wavelength tunable Dy^{3+}/Er^{3+} co-doped ZBLAN fiber lasers. Optics Express. 2021;**29**: 38646. DOI: 10.1364/oe.443808

[19] Bharathan G, Woodward RI, Ams M, Hudson DD, Jackson SD, Fuerbach A. Direct inscription of Bragg gratings into coated fluoride fibers for widely tunable and robust mid-infrared lasers. Optics Express. 2017;**25**:30013. DOI: 10.1364/oe.25.030013

[20] Hansen SG, Robitaille TE. Formation of polymer films by pulsed laser evaporation. Applied Physics Letters. 1988;**52**:81-83. DOI: 10.1063/1.99332

[21] Goya K, Sasanuma H, Ishida G, Uehara H, Tokita S. Fusion splicing of plastic optical fibers using a mid-IR fiber laser. Applied Physics Express. 2023;**16**: 052006, 1-8. DOI: 10.35848/1882-0786/acd491

[22] Industrial Applications Archives – Femtum. n.d. Available from: https://femtum.com/category/industrial-applications/ [Accessed: April 24, 2025]

[23] Bernier M, Faucher D, Vallée R, Saliminia A, Androz G, Sheng Y, et al. Bragg gratings photoinduced in ZBLAN fibers by femtosecond pulses at 800 nm. Optics Letters. 2007;**32**:454. DOI: 10.1364/ol.32.000454

[24] Bérubé J-P, Bernier M, Vallée R. Femtosecond laser-induced refractive index modifications in fluoride glass. Optical Materials Express. 2013;**3**:598. DOI: 10.1364/ome.3.000598

[25] Bernier M, Trépanier F, Carrier J, Vallée R. Efficient writing of bragg gratings through the coating of various optical fibers. In: Bragg Gratings, Photosensitivity, and Poling in Glass Waveguides, BGPP 2014. Washington, D.C., USA: Optica Publishing Group; 2014. pp. 3-4. DOI: 10.1364/bgpp.2014.bm2d.3

[26] Goya K, Matsukuma H, Uehara H, Hattori S, Schäfer C, Konishi D, et al. Plane-by-plane femtosecond laser inscription of first-order fiber Bragg gratings in fluoride glass fiber for in situ monitoring of lasing evolution. Optics Express. 2018;**26**:33305. DOI: 10.1364/oe.26.033305

[27] Fernandez TT, Xu L, Gross S, Bharathan G, Ams M, Privat K, et al. Physical mechanisms of femtosecond laser induced refractive index change in direct-written mid-infrared fiber Bragg gratings. APL Photonics. 2024;**9**:110802. DOI: 10.1063/5.0224504

[28] Liu R, Luo Z, Wu L, Ran J, Wang Y, Lv C, et al. Efficient inscription of high quality FBGs on ZBLAN fiber with a femtosecond laser. Journal of Lightwave Technology. 2025;**43**:4429-4434. DOI: 10.1109/JLT.2025.3531007

[29] Goya K, Uehara H, Konishi D, Murakami M, Tokita S. High-index-contrast Bragg gratings fabricated in fluoride fiber with 513-nm femtosecond laser. In: Proceedings Volume 10899, Components and Packaging for Laser Systems V. Vol. 10899. Washington, D. C., USA: SPIE; 2019. pp. 143-148. DOI: 10.1117/12.2508362

[30] Oda K, Konishi D, Misawa A, Sahara R, Tsuzuki S, Murakami M, et al. Laser Processing Technology with Mid-infrared Er Fiber Laser. Hikari Alliance (Optronics Alliance). Tokyo, Japan: Japan Industrial Publishing Co, Ltd; 2021. p. 4 (in Japanese)

[31] Misawa A, Oda K, Maeda K, Konishi D, Murakami M. Residual stress control in mid-infrared laser rounding of glass plate edges by reciprocal scanning assist laser irradiation. In: The Proceedings of Mechanical Engineering Congress, Japan 2022. Tokyo, Japan: The Japan Society of Mechanical Engineers; 2022. pp. J131-06. DOI: 10.1299/JSMEMECJ.2022.J131-06

[32] Yao W, Uehara H, Uehara H, Kawase H, Chen H, Yasuhara R, et al. Highly efficient Er:YAP laser with 6.9 W of output power at 2920 nm. Optics Express. 2020;**28**:19000-19007. DOI: 10.1364/OE.395802

[33] Uehara H, Yasuhara R, Goya K, Tokita S. High power solid-state and fiber lasers at 3-4 μm wavelength and their applications. In: 16th Pacific Rim Conference on Lasers and Electro-Optics (CLEO-PR 2024), Incheon, Korea. Washington, D.C., USA: Optica Publishing Group; 2024

[34] Uehara H. Development of Mid-Infrared Solid-State Lasers and Their Applications for Processing and Remote Sensing. Tokyo, Japan: Kogaku; 2025. p. 54 (in Japanese)

[35] Schäfer CA, Uehara H, Konishi D, Hattori S, Matsukuma H, Murakami M, et al. Fluoride-fiber-based side-pump coupler for high-power fiber lasers at 2.8 μm. Optics Letters. 2018;**43**:2340. DOI: 10.1364/ol.43.002340

[36] Withford MJ, Jackson SD, Williams RJ, Hudson DD. Single-frequency fiber laser operating at 2.9 μm. Optics Letters. 2013;**38**(14):2388-2390. DOI: 10.1364/OL.38.002388

Chapter 8

Laser Facial Modulation after Cheiloplasty: *The Teixeira-Pinheiro Technique (2025)*

Heitor Teixeira, Ruben Pinheiro and Carolina Teixeira

Abstract

This chapter explores the *Teixeira-Pinheiro Technique (2025)*, an innovative method employing picosecond Nd: YAG lasers for the remodelling of post-surgical hypertrophic scars which remains a frequent challenge in patients undergoing corrective cheiloplasty for cleft lip deformities. The chapter details the methods of a prospective observational study involving over thirty patients, all treated with picosecond Nd: YAG lasers. The results show the marked clinical and aesthetic aspect improvements in all participants of the treated scars, with minimal adverse effects, underscoring the picosecond laser's potential in inducing controlled photomechanical damage, ideal for delicate facial areas. The discussion delves into the technique's safety and efficacy, and proposes the integration of non-ablative biomodulatory lasers in the early postoperative phase for even more robust scar prevention. We conclude that the Teixeira-Pinheiro Technique (2025) establishes laser therapy as a first-line treatment for surgical and post-traumatic scars or contractures, highlighting the picosecond Nd: YAG laser. The chapter also discusses its low morbidity, minimal damage to superficial skin layers, and the significant functional and aesthetic benefits observed. We reaffirm the need for future research to validate and optimise these approaches as laser technology continues to evolve, promising even greater advancements in patient aesthetics and health.

Keywords: laser therapy, picosecond Nd: YAG, cleft lip, postoperative hypertrophic scarring, scar remodelling, *Teixeira-Pinheiro technique 2025*

1. Introduction

Cleft lip, with or without cleft palate (CLP), is among the most frequent and visible congenital malformations of the craniofacial complex, affecting approximately one in every 700 live births globally, with regional prevalence ranging from 0.5 to over 2 per 1000 live births depending on ethnicity, geography, and environmental exposures [1, 2]. These anomalies originate during embryogenesis—specifically between the 4th and 7th weeks of gestation—when the developing facial prominences fail to fuse properly. More specifically, cleft lip results from the incomplete fusion of the medial nasal and maxillary processes, while cleft palate arises from the failure of palatal shelves to elevate and merge including the nasal septum [3]. The resulting deformities may manifest as unilateral or bilateral clefts, affecting only the lip, or extending through the alveolus and palate, disrupting both aesthetic form and orofacial function.

The causes of CLP are multifactorial, involving a complex interplay of genetic and environmental factors. Numerous genes have been implicated in lip and palate formation, including *IRF6*, *MSX1*, *PVRL1*, and *TGFB3*—all of which regulate epithelial-mesenchymal interactions, tissue adhesion, and programmed cell death critical to craniofacial development [4, 5]. Environmental teratogens such as maternal smoking, alcohol use, anticonvulsant medications, and folate deficiency during early pregnancy significantly increase the risk of cleft formation. Maternal smoking alone is associated with up to a 20–30% increased risk of orofacial clefts in offspring [6]. Despite progress in genetic mapping and population surveillance, no single etiological pathway explains all cases, and variability in phenotype and severity remains high, even within the same population or family group.

The clinical burden of CLP extends far beyond the initial diagnosis. These anomalies compromise several fundamental functions, including feeding, breathing, speech, hearing, and facial expression. Additionally, the psychological and social ramifications—ranging from reduced self-esteem and bullying to parental anxiety and stigma—persist throughout childhood and adolescence [7, 8]. The upper lip is central not only to oral competence and phonation but also to emotional expression and identity; thus, achieving a natural, symmetrical, and mobile lip is essential for both function and psychosocial development.

The management of children with CLP is necessarily multidisciplinary, and increasingly transdisciplinary, involving close collaboration among specialists in plastic, paediatric and maxillofacial surgery, otolaryngology, speech-language pathology, orthodontics, audiology, genetics, psychology, paediatrics, dermatologists, and nursing. This team-based approach ensures that the patient's complex and evolving needs—from infancy to adulthood—are addressed holistically. For instance, early feeding strategies and speech interventions must be coordinated with surgical timing, while orthodontic and dental care often spans multiple stages of craniofacial development. Equally important are the contributions of psychosocial and family support services, which play a key role in ensuring adherence to care plans and long-term well-being.

The most successful cleft care programs are those that not only integrate these disciplines but promote continuous communication and joint decision-making among professionals and families alike.

Surgical correction of the cleft lip—known as cheiloplasty—is typically performed after 3 months of age. The goals of cheiloplasty extend beyond simple closure of the defect. Surgeons aim to reconstruct a structurally sound and aesthetically balanced lip, restore the continuity of the orbicularis oris muscle, align the vermilion border, recreate the Cupid's bow, and preserve or restore the nasal sill. Over the decades, numerous surgical techniques have been developed to achieve these objectives, including the Millard rotation-advancement, Tennison-Randall triangular flap, and Rose-Thompson straight-line methods. However, contemporary cleft care increasingly emphasises not only anatomical correction but also functional and dynamic restoration of the lip.

In our clinical experience, we base our technique on the Sommerlad anatomical approach, which prioritises the complete dissection, mobilisation, and reapproximating of the orbicularis oris muscle across the cleft gap [9]. This foundational step re-establishes muscular continuity and tension vectors, allowing for better long-term function and minimising residual deformity. Moreover, the proper reconfiguration of nasal musculature and alar base is vital in reducing nasal asymmetry, a common challenge in unilateral cleft repairs. To improve the aesthetic results and to break the vertical scar line, there is an incorporated small triangular flap modification, inspired by the Fisher anatomical subunit approximation technique [9]. This additional geometric element enables more natural philtre column reconstruction and aligns skin units based on topographical subunits rather than arbitrary incisions.

The muscular realignment allows for a more harmonious outcome—both statically and during facial movement.

Despite technical precision, post-surgical scarring remains a persistent and sometimes unpredictable issue. Scar quality varies widely based on individual healing response, skin type, age, wound tension, and post-operative care. In many cases, the scar may evolve into a hypertrophic, hyperpigmented, or contracted formation—particularly in patients with a strong fibroproliferative response or delayed epithelialization.

Given the prominent location of the upper lip, even a moderately conspicuous scar may draw disproportionate attention, compounding the psychosocial burden of the cleft. Scar contracture may also tether the lip, restrict animation, or distort nasal form, necessitating further revision surgeries.

Our post-cheiloplasty scar care protocol integrates early, comprehensive measures. Parents are instructed in gentle lip massage for desensitisation and mobilisation, aiming to improve collagen alignment and tissue softening. This is initiated weeks post-operatively and continued daily for several months, alongside topical silicone-based products and sun protection. While these conservative practices are foundational, they may not always ensure optimal cosmetic or functional outcomes, especially in complex cases. Crucially, though scar massage aids desensitisation, its efficacy in improving scar appearance remains debated, with recent studies suggesting surgical technique and suture material have a more significant impact on scar quality [10–12].

Orthodontic treatment is a cornerstone in the multidisciplinary management of children with cleft lip and/or palate, addressing the critical aspects of craniofacial growth and function [13].

These patients often present complex dental and skeletal changes requiring early corrective intervention, ideally starting during deciduous or mixed dentition. Early intervention facilitates a more interdisciplinary approach, integrating functional orthodontics with adjuvant therapies such as speech therapy, paediatrics, and maxillofacial surgery. Beyond aesthetic and social impacts, patients may experience issues with sucking, lip sealing, diction, and hearing.

Early orthodontic intervention is particularly crucial when patients present with specific alterations, including:

- *Transverse*: anterior or posterior crossbite.
- *Vertical*: open bite or deep bite.
- *Sagittal*: maxillary/mandibular protrusion or retrusion.
- *Dental malocclusion*.
- *Need for reconstructive surgery*.

The use of transparent aligners has significantly enhanced comfort for these patients. Aligners allow independent removal for eating and oral hygiene, requiring only dental support *via* custom manufacturing and attachments. Their independence from other oral structures permits simultaneous adjuvant therapies like dental laser.

Worn for 22 hours daily, aligners eliminate comorbidities associated with traditional fixed appliances, previously the primary orthodontic intervention. (**Figures 1** and **2**).

Furthermore, the aligners are translucent and therefore practically imperceptible.

The reduction in friction at the level of appliances represents an added advantage for patients who have been subject to other types of treatment and intraoral therapies for a long period of time.

Meanwhile, in the vast majority of cases, treatments are multifactorial and involve intervention not only at the dental level, but also at the musculoskeletal and surgical level, this means that they are treatments that last for several years.

Therefore, the more comfortable and aesthetic they are for the patient, the greater their compliance will be and, consequently, the better the result obtained. (**Figures 3** and **4**).

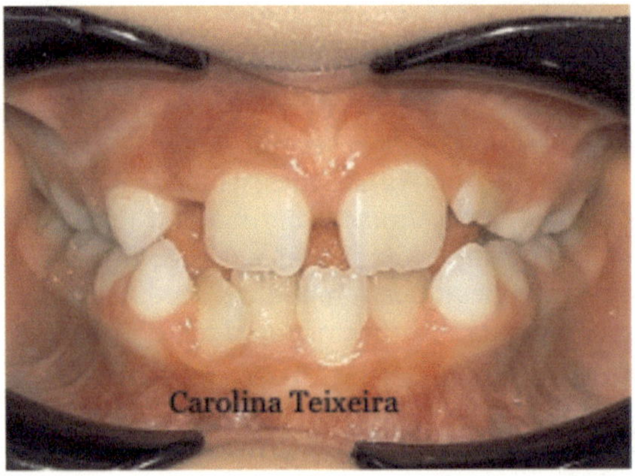

Figure 1.
Before orthodontic treatment.

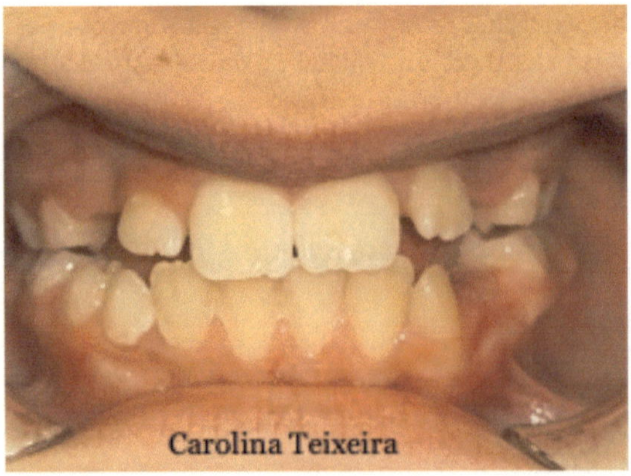

Figure 2.
After orthodontic treatment.

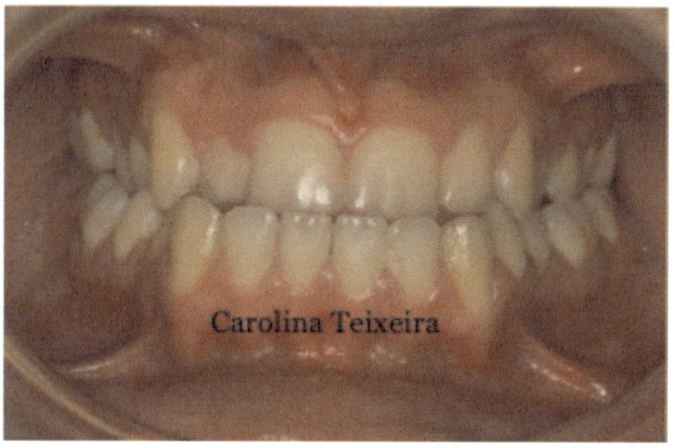

Figure 3.
Before using aligners.

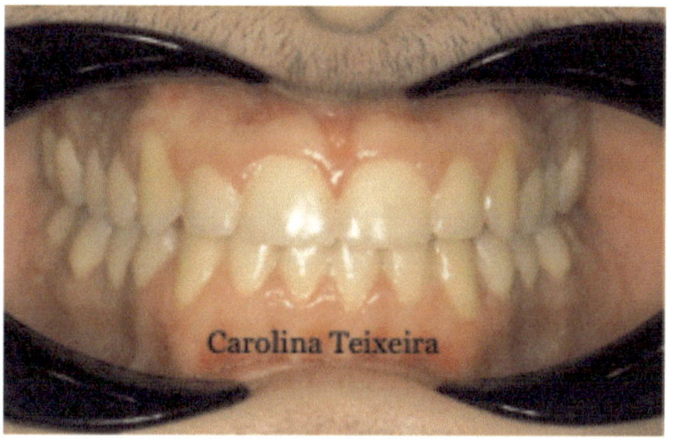

Figure 4.
After using aligners.

Dental orthodontic with clear aligners in patients with cleft lip or palate [14] represents the state of the art in terms of cutting-edge orthodontic treatments indicated for intervention in this niche of patients (**Figures 3** and **4**). However, the literature lacks clinical studies in the form of randomised clinical trials on the patients' perspective and perception of the results of orthodontic treatments in this specific group of patients, especially with this type of intraoral devices.

Within this complex context, the role of non-invasive modalities for scar modulation becomes increasingly significant.

In particular, the evolution of laser therapy—especially with the advent of *picosecond laser*—has introduced new opportunities to alter scar architecture, pigmentation, and pliability with minimal morbidity.

Since the discovery of the Laser (1960), various types have been used in aesthetic and reconstructive medicine for over 40 years. Initially met with scepticism, laser treatment has proven increasingly effective and elective.

This chapter discusses the indications and limitations of *picosecond laser* application in post-cleft lip, palate, or lip-palatine cheiloplasty treatment, drawing on the published literature and extensive personal experience. The novel *picosecond laser* is a non-invasive, safe, and effective method with minimal complications, offering superior results compared to traditional soft tissue augmentation or surgery [15]. Its key differentiator is a non-invasive approach with minimal transient erythema, discomfort, or pain, achieved by increasing collagen, tightening tissue, reshaping the lip, and renewing surrounding skin to create functional new tissue and a new smile [16] and it is notable for its 10^{-12} or 1/1000,000,000,000 of a *second* pulse duration.

Effective treatment requires understanding how visible scars and functional issues (e.g., phonetic impairment) impact a patient's family and social life, considering the child's orofacial development, family context, fears, and stigma.

The skin, our largest organ and protective barrier, is central to this, relying on collagen for firmness and elastin for elasticity.

Scars are fibrous tissue resulting from skin damage due to wounds, cuts, or surgery (**Table 1**).

In addition to the aesthetic impact, scars can affect the mobility of tissues and the function of the skin in the affected area, and may even compromise organs in more severe cases.

TIPE of SCARS and their CHARACTERISTICS	
• **Normal scars**	a. They are flat and have the same colour of the skin 
• **Skin discoloration and epidermal irregularities**	b. They do not harm one's health, but they are unpleasant for those who have them 
• **Atrophic scars**	c. They tend to be concave and may give the skin a thinner appearance 
• **Hypertrophic scars**	d. They are more raised, thick and reddish, and may present either hyperpigmentation or hypopigmentation 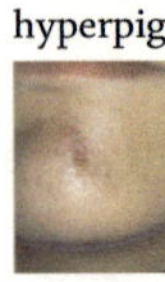
• **Keloids**	e. They are a type of hypertrophic scar that extends beyond the original wound margins, they can be large, hard, raised often cause intense pruritus or itchy. 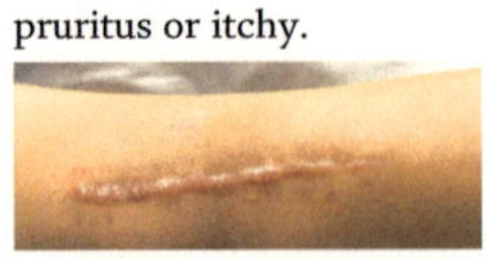
• **Contractures**	f. They restrict movement by sticking the skin and underlying tissue together during healing 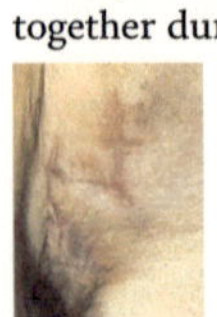

Table 1.
Type of scars.

The *scar* formation process involves three main phases: *inflammation*, *proliferation* (formation of new tissue and blood vessels), and *maturation* (tissue remodelling).

If the *scar* causes physical or aesthetic discomfort, there are treatments that range from ointments and creams to medical procedures, the most appropriate and elective being the application of laser, which induces skin retraction, reducing thickness and eliminating collagen and elastin fibbers and at the same time, it has a biostimulator effect on new collagen production and elastin.

Each organism's responses are different, but ideally, treatment for *scars* should begin after the first week of surgery, if the area is not open or infected.

2. Materials and methods

2.1 Study design

We have selected 30 cases for laser facial modulation after cheiloplasty.

This is a preliminary retrospective observational study of this new technique which we have called *Teixeira-Pinheiro 2025.*

The basic criteria were cross-sectional, and we have patients among 8 months until 20 years.

Twenty of them were male and 10 were female.

This study lasted from 24 of May 2023 until 6 July 2025.

The follow-up is under control every 2 months initially, but many of them were seen every 3 months.

Our patients come exclusively from our private office.

The first question that guided the planning of the treatment for the elimination of scars after cheiloplasty had to do with the choice of the type of laser to be used.

To make this decision, the knowledge acquired over decades in the aesthetic area in the elimination of wrinkles, stretch marks, and tattoos allowed us to acquire a good understanding of the skin and especially of scars, especially in their remodelling and re-epithelialization phases, which have been achieved for many years [17].

The aim of significantly improving the effects that the different lasers cause in this, let us to the primary selection of the most recent *Picosecond Laser* to which we associated in 4 cases the *CO_2 Laser* [18].

In the total group, we used only a Picosecond Nd: Yag 532 nm Laser, spot size of 4, 3 mm mainly in the total scar, and 2 mm on the contours of it.

The frequency was initially at 6 Hz at 4, 3 mm spot size and 2 on 2 mm spot size.

So, the power was from 223,6 mJ till 278,5 mJ respectively.

2.2 Characteristics

A *picosecond* is an *SI unit* of time equal to 10–12 or 1/1000,000,000,000 of a *second*. That is one trillionth, or one millionth of one millionth of a *second*, or 0.000000000001 *seconds.*

Based on previous experience using *picosecond lasers* and their effects on the skin, we postulate the hypothesis of achieving a marked regression of the scar fibrotic component after cheiloplasty [17].

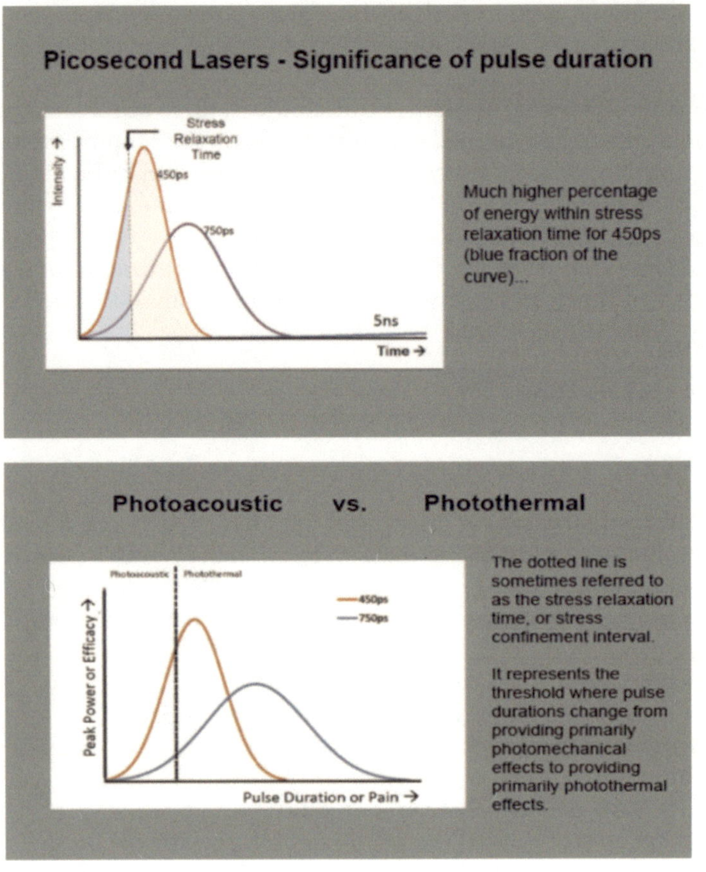

Figure 5.
Pulsed duration and effect.

Its ability to obtain a photoacoustic effect (**Figure 5**) induces the introduction into the skin of a high energy density that enables the destruction of elastin fibbers and the breakdown of fibrotic tissue into very fine particles that will be subject to high metabolic reabsorption, which are then quickly cleaned and excreted from the body by the macrophage system (**Figure 6**).

The knowledge of these essential characteristics allowed us to define and adapt a *Teixeira-Pinheiro 2025* protocol that we applied to our patients who participated in this study, which consists of the following step:

2.2.1 Diagnosis

Detailed clinical history.
Skin cleansing and preparation.
Photographic record.

2.2.2 Technique

Anaesthesia and pain reduction—the discomfort associated with the ablative application of the laser, in this case in young children, requires a topical anaesthetic

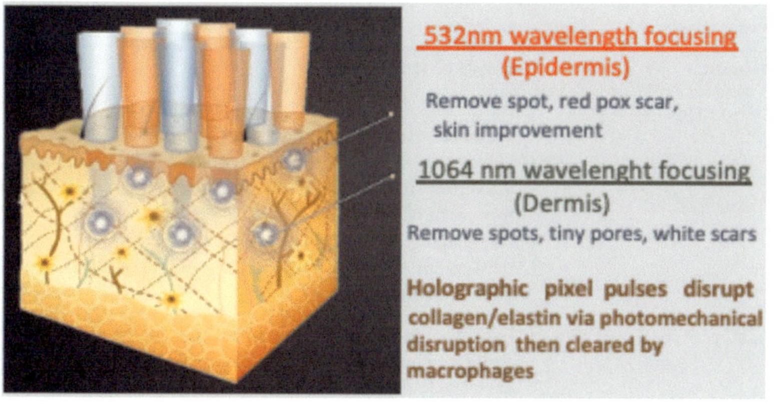

Figure 6.
Disrupt collagen and elastin.

strategy with the application of a compounded ointment of *30% lidocaine + 5% tetracaine + 5% prilocaine*, which is used beforehand around the peri-scar area targeted by the laser, to which is added, after 20 minutes of action, a subcutaneous injection of *mepivacaine hydrochloride 36 mg + adrenaline tartrate 0.032 mg in 1.8 ml* of injectable solution, in the various planes where the intervention was planned and, if necessary and indicated, also a *sensory nerve block* of the infraorbital nerve.

Children with a clear component of anxiety, or who have very protective or disturbed parents, should also use *midazolam hydrochloride oral solution* in a body dose appropriate to their age and administered 30 minutes before the intervention.

So, we start treatment by *picoseconds laser* wavelength primarily 532 nm, spot size of 4 mm, 3 mm or 2 mm, frequency rate of 4 hz or 1 hz when needed, with *E.D. energy density* from 230 till 279,5 mJ, must be then applied, as we detailed described in clinical cases.

2.2.3 Session

Despite the fact that each case is different and each patient reacts in an individual way, after these years of treatments, it was possible to obtain a practically standard pattern for the treatment sequence.

Initially, the first two sessions are monthly and the following ones every two months, up to six months and then every three months.

2.2.4 Recommendations after laser treatment

Topical broad-spectrum antibacterial ointment is applied immediately following the procedure and a prescription of an antibacterial ointment for exclusive application to the dermis twice a day for 5 days, for no longer than that time, to prevent maceration of the thin scar crust, followed by the indication of *digital pressure massage twice* a day (**Table 2**).

As in all clinical practice, only proper observation and careful observation of the evolution, whether it is desirable or not, allows us to obtain an assertive diagnosis that provides us with good treatment, good practice, and, above all, the desired cure.

Example of post-operative recommendations	
Patient's doubt	**Example of the answer**
What should I expect after laser treatment?	• the treated area may become pink and swollen • the swelling may last up to 7 days or more • some patients may experience redness for 7 days or more • discomfort may occur, rarely bleeding or bruising also later the formation of scar scabs • skin peeling may occur between 3 and 5 days • total postoperative healing varies between 15 and 20 days, depending on each one
How should I care for the treated area?	• Initially abstain water to avoid moistening the treated area • Apply the prescribed antibiotic ointment twice for five days exclusively on the skin • Eschew using any antiseptic or occlusive dressing • After the skin has completely healed, perform a digital pressure massage on the area daily
What should I must abstain?	• touching or rubbing the treated skin • directly contact with pets until the skin is healed • prolonged direct sun exposure • using sunscreen on the treated skin

Table 2.
Post-operative recommendations.

Treating patients with laser means, on the one hand, that one is qualified to perform a role for which thousands of hours of study have been spent, both in theoretical knowledge of its foundations and its distinct characteristics, and on the other, that one has a unique tool that allows one to resolve, in an instant, both acute and chronic diseases that would not be curable by any other therapeutic means.

Patient satisfaction is intrinsically linked to the establishment of realistic expectations and ongoing clinical follow-up during the rehabilitation period. Inadequate monitoring may result in missed opportunities for diagnosing potential complications or infections, although such occurrences are exceedingly rare. It is essential that patients do not feel abandoned throughout the recovery process, which can vary in duration depending on the type of ablation procedure performed.

Thus, in the following clinical cases we will observe the application and evolution of a laser treatment, with the presentation of before and after treatment; we chose not to dwell on excessive comments as we consider that the evidence presented is clearly enlightening as to what the choice and application of the laser, or lasers, appropriate to the target of our therapy consists of, depending on the case.

The mandatory selection of cases only reflects a very restricted part of the universe of patients treated, but all clinical cases are duly documented, are part of our vast and differentiated experience, and faithfully translate the main objective of the *creation*, *development*, and *differentiation* of the presentation of this new innovative *technique*.

3. Clinical cases

See **Figure 7**.

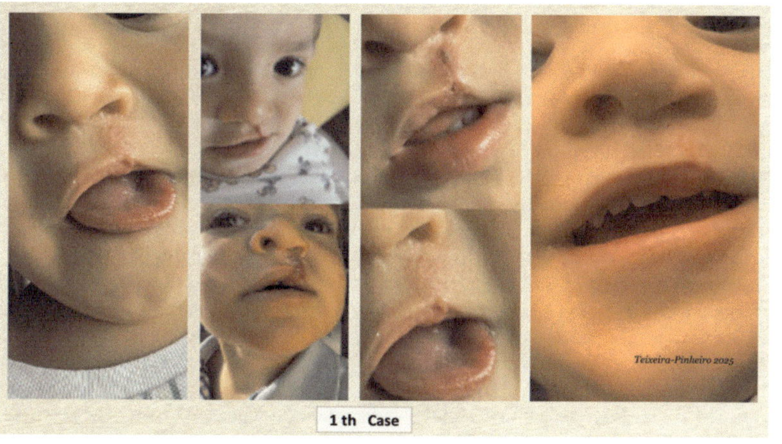

Figure 7.
1th case: Cleft lip with internal fold and hypertrophic skin scar.

1_{Th} case

We present in these figures the evolution of a treatment of a child who starts this laser care with one's years old.

He stars its first lip surgery with 3 months and have them next surgery of palate at 9 months.

We can appreciate in this figures the evolution from the beginning, after 3 days, then 10 days passed reporting in the same figure with a comparison of days after the first application, and finally at three months.

Note that he only had three treatments, the initial one, the second one at 2 months, and the third and last one three months later.

The treatment was performed with a *picoseconds laser* issuing *with* a wavelength of 532 nm, a 3-mm spot in the first pass, and 2-mm spot in the second pass, which was applied only to the hypertrophic dermis.

The frequency used was always at 4 Hz.

It should be noted that the background of the treatment occurred in a short space of 5 months.

We highlight not only the reverse of the scars, but also the excellent aesthetic, result achieved.

See **Figure 8**.

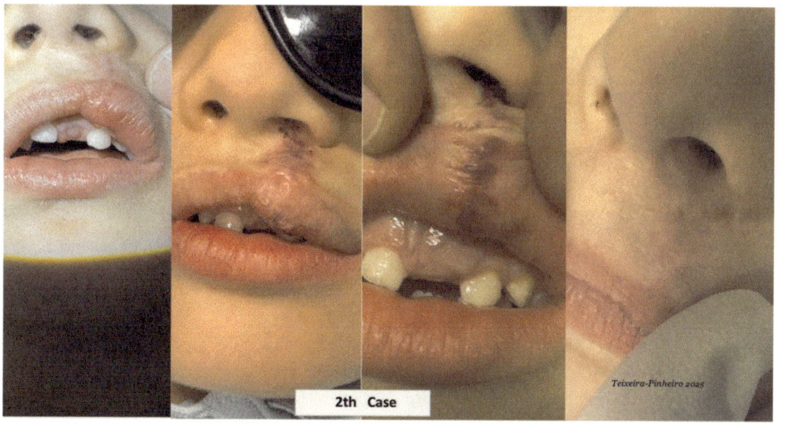

Figure 8.
2th case: High hypertrophic skin scar with intense reddish excrescence.

2$_{Th}$ case

This clinical case reports to a 6 years old child, who performs is first lip's surgery at the three months of age and then the palate's at the 5 years of age.

He presents a high hypertrophic scar between the upper lip and the basis of the nose and an intense reddish colour excrescence, specifically on the innermost fold of the upper lip.

The lips are the component of the oral cavity that synthesises more functions and emotions.

Non-verbal communication is established by the body, all movement, gesture, posture, and facial expression that shows a willingness to communicate directly or subtly, consciously or unconsciously. The smile is one of the expressions that most appeal to us, for simplicity and especially for your puzzle.

Figure 335 shows the effect of the laser in reducing lip over thickening excrescence immediately after one passage performed with a *picoseconds laser* issuing *with* a wavelength of 532 nm, with a 5-mm spot in the first pass, who allows us to be deeper in the target and then a 3-mm spot in the second pass.

A 3-mm spot in the first pass and 2-mm spot in the second pass were applied only to the hypertrophic dermis of the upper lip.

The frequency used was always at 4 Hz and the energy density deposited various between 212,5 and 279,5 mJ.

The patient performed only three treatments at the third month and then at the seven month after starting.

It is remarkable not only the immediate reduction in the thickness of the labial excrescence, but also the almost total reduction of the demand scar that it presented highlights not only the reverse of the scars, but also the excellent aesthetic result achieved.

See **Figure 9**.

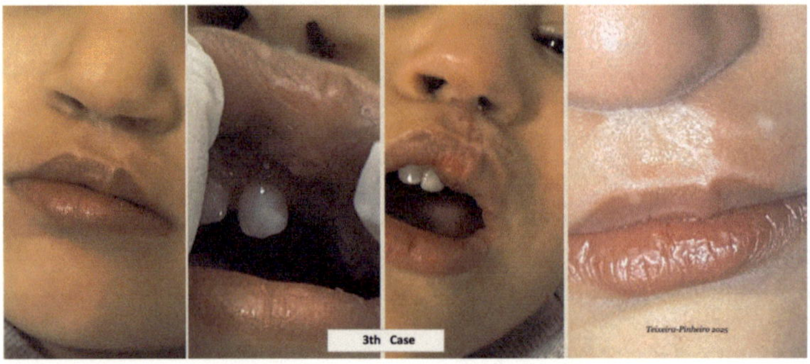

Figure 9.
3th case: Hypertrophic skin scar with reduction of the dermal scar.

3$_{Th}$ Case

The patient of this case was 2 years old child, who was submitted to palatoplasty at 3 months, then after a new surgery at 9 months with tubes in the ear and when he comes for consultation for the first time he has been made frenectomy 11 days before.

He presented hypertrophic scars in a network with different thicknesses and orientations extending to the inner face of the upper lip where they were very contracted and a noted dyschromia on the right side of the upper lip.

As we can see in Figure 24 that shows us the effect of the laser in reducing lip over thickening excrescence immediately after the first passage of the *picoseconds laser* issuing *with* a wavelength of 532 nm, with a 4-mm spot, who allows us to be more deeper in the target and then a 3-mm spot in the second pass.

On the hypertrophic dermis, we first apply the wavelength of 1064 nm with the spot of 5 mm to go deeper in the dermis and immediately after we took over the 532 nm 3-mm spot in the first pass and 2-mm spot in the second pass, so we made three different passages.

The same protocol was used in the sequential treatments that occurred 2, 4, 9, and 11 months later.

If we closely examine detail the scars have regressed significantly, particularly the one on the upper lip, and the dyschromia that existed on the right side of that lip has also diminished considerably.

Despite the fact that the child has grown over a period of four years—during which the scars would be expected to increase in size proportional to facial growth—a marked reduction in their size was instead observed.

See **Figure 10**.

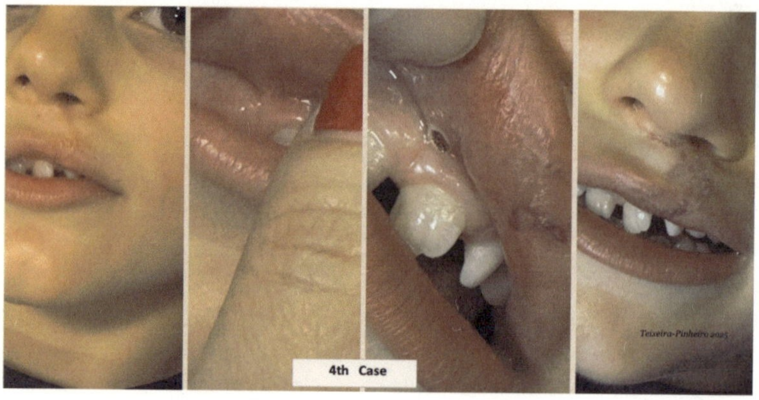

Figure 10.
4th case: Hypertrophic scar after frenectomy.

4_{Th} case

This patient, an 8-year-old child, performs his first lip's surgery at the age of 8 months and recently, three months ago he underwent a frenectomy which left him with a hard contractile scar.

Also, he presents a hypertrophic scar between the upper lip and the basis of the nose.

Figure 25 shows the effect of the laser in reducing lip over thickening excrescence immediately after one passage performed with a *picoseconds laser* issuing *with* a wavelength of 532 nm, with a 5-mm spot in the first pass, who allows us to be deeper in the target. and then a 3-mm spot in the second pass.

Then a 2-mm spot in the third pass was applied mainly only into the centre of the hypertrophic dermis of the upper lip.

Regarding the hard contractile scar after frenectomy, it was decided to use the CO_2 laser first in ultra-pulsed mode with a 1-mm spot and an D.E. (energy density) of 2 joules with a defocusing of approximately 4 cm to obtain volatilisation of the target and then in the most central area, there in continuous mode and with 1 joule, a carbonisation of the central nucleus was performed.

This is an excellent example how different lasers can be applied on the target so that we can achieve our goal.

We note not only the immediate reduction in the thickness of the hard and contractile scar after frenectomy, but also the performance of labial excrescence, which allows the patients to begin to express himself orally much better.

See **Figure 11**.

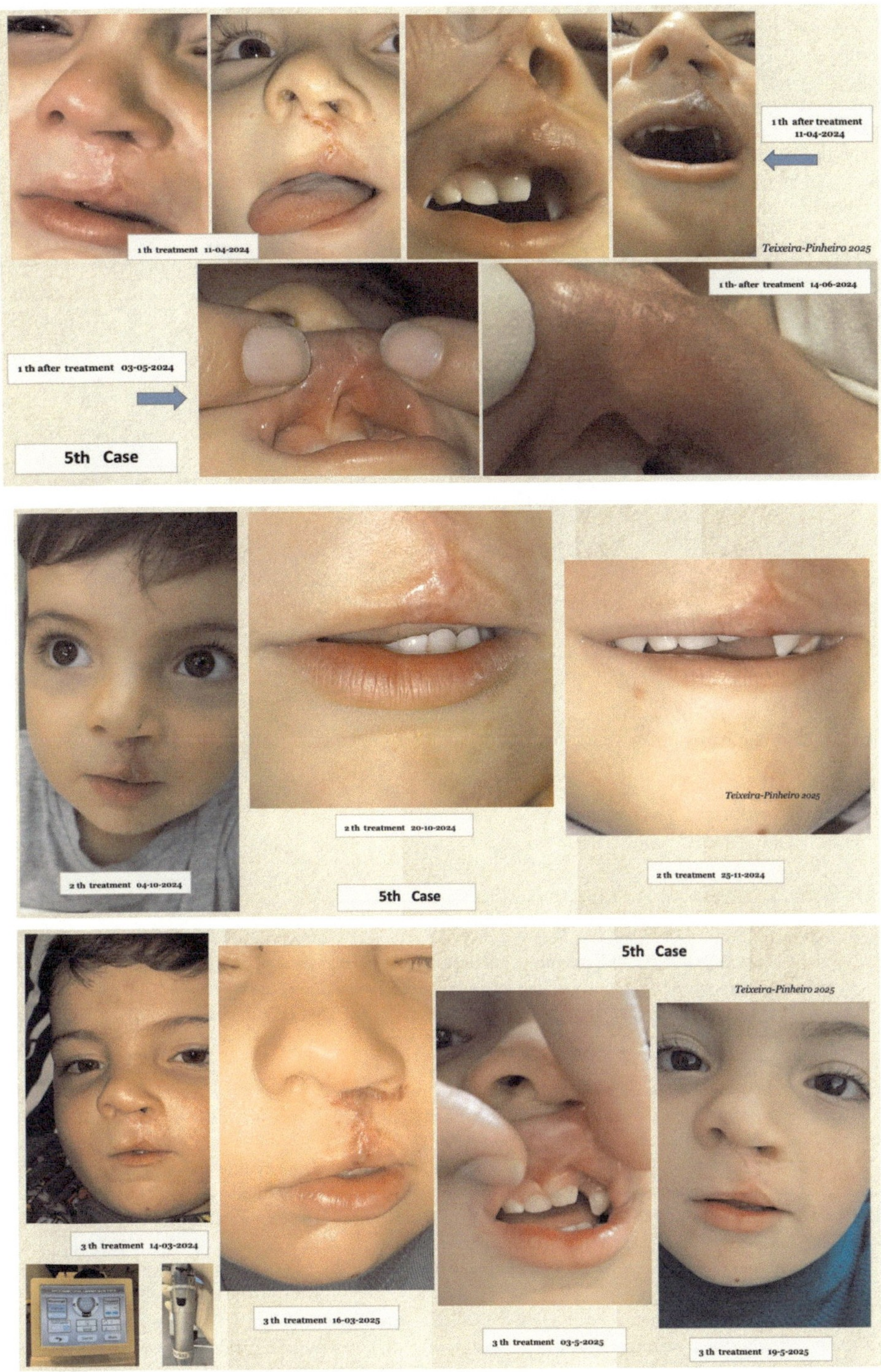

Figure 11.
5th case: Hypertrophic scar after cheiloplasty.

5Th case

The sequence of figures presented in this case documents the detailed follow-up of three therapeutic interventions performed on a paediatric patient following initial surgical correction of the palate, cleft lip, and excision of a neoplastic nasopharyngeal tumour.

The patient underwent the first session of picoseconds laser therapy three years and three months after the initial surgery starting with picoseconds laser on the left supralabial hypertrophic scar and also inner the per labial excrescence of the aforementioned lip.

So these dates refer to the evolution obtained with the perfection acquired, post-laser.

In these treatments, we choose to always apply the wavelength of 532 nm with two sequential passes.

Thus, for the first pass we use a 3-mm spot and an energy density of 254.6 mJ.

Then, a 2-mm spot in the second pass was applied mainly only into the centre of the hypertrophic dermis of the upper lip.

As we can clearly see thickening excrescence immediately almost disappear two months after the first treatment and one year passed we cannot almost distinguish any trace of it.

Regarding the hypertrophic scar, there is a clear marked decrease in its thickness diameter and skin tone.

The follow-up will now occur and extended intervals, with each future assessment scheduled every six months as a part of more extended periodicity needed.

See **Figure 12**.

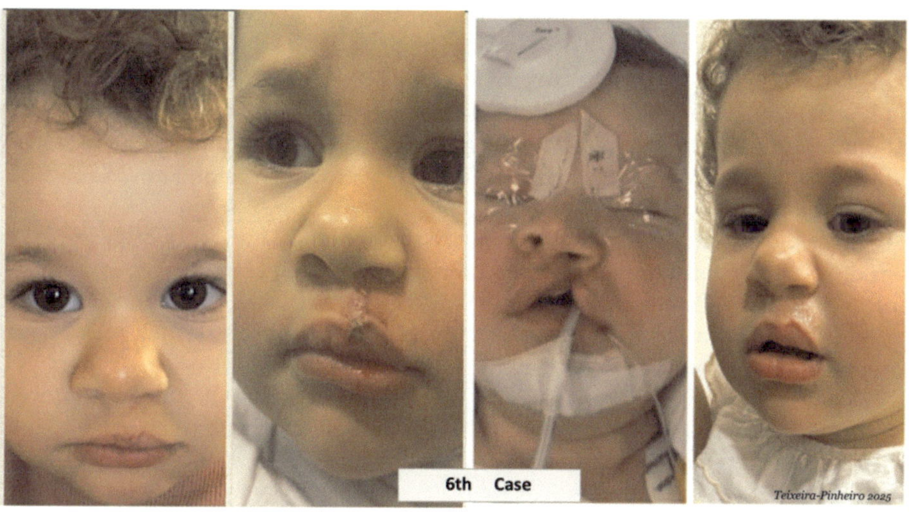

Figure 12.
6th case: Cheiloplasty scars and picoseconds Laser.

6Th case

Figure 12 referes a patient who was submitted to left lip surgery (also called *cheiloplasty*) which is a procedure performed to correct a cleft or opening in the upper lip that may have been present since birth (congenital malformation). This condition occurs when the tissues that form the lip do not fully join together during foetal development.

The objectives of surgery are to close the cleft lip; to improve function (speech, eating and breathing); to restore the aesthetic appearance of the face; and to promote proper development of the mouth and face.

The surgeon reconstructs the lip, joining the muscles and tissues symmetrically.

Fundamental for the development of the child, as the aesthetic aspect is also increasingly decisive in the child's interpersonal involvement.

> Therefore, the addition of another medical tool to complement it, the *picoseconds Laser* created a superior synergy that become like this the *gold standard* we were looking for in this appearance and that this figures clearly show us.

4. Discussion

Research into effective dermo-aesthetic facial remodelling treatments has led to the widespread adoption of laser technology in dermatology and dermatofunctional physiotherapy [19]. Laser facial modulation, a key therapeutic option, precisely controls electromagnetic radiation parameters (wavelength, fluence, energy density, pulse duration to stimulate tissue biostimulation, extracellular matrix regeneration, mitochondrial activity, and microcirculation), ultimately promoting tissue regeneration and optimal scarring in dermal lesions [20, 21].

Conversely, laser ablation therapy achieves tissue destruction by rapidly elevating the target tissue's temperature, inducing protein denaturation. Crucially, successful ablation requires precise laser energy delivery and sufficient thermal energy to destroy target tissue safely, while minimising collateral thermal damage to surrounding healthy neurovascular structures [22, 23]. The varying absorption of light by cells, dependent on photon wavelength (e.g., near-infrared by cell walls, red by mitochondria, UV by the nucleus), triggers photochemical reactions [20, 21].

This leads to the formation of reactive oxygen species (ROS) in mitochondria; while lower ROS levels enhance cellular activity, high concentrations are toxic, capable of eliminating old cells and stimulating the regeneration of new tissue [22]. Despite the clinical success of ablative laser procedures, thermal diffusion can still cause damage to healthy biological tissues [23].

Therefore, we shall briefly revisited some fundamental concepts of laser physics that underpin the treatments and served the basis for the creation of this novel technique "*Teixeira-Pinheiro 2025*," based on the clinical cases presented, the principal differences outlined below- begin with the type of laser emitter, which may operate in either continuous or pulsed mode, each exhibiting distinct biophysical properties [24].

The *continuous emission* lasers (*CW*)—**Figure 13**—are used in venous photocoagulation, in the endoluminal treatment—CELIV, in photodynamic therapy-PDT, *pulsed emission* in nonsurgical therapy laser—NSL and in the diagnose with the low-power laser—**Figure 13**.

The *super-pulsed emission* lasers—**Figure 14**—and *Ultra-pulsed*—**Figure 15**—(nanotech, Qswitched) are very powerful with very short pulses / Q-switch that cause

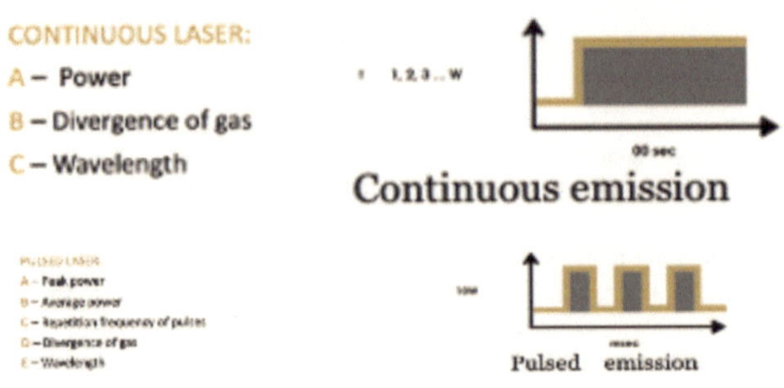

Figure 13.
Continuous emission and pulsed emission.

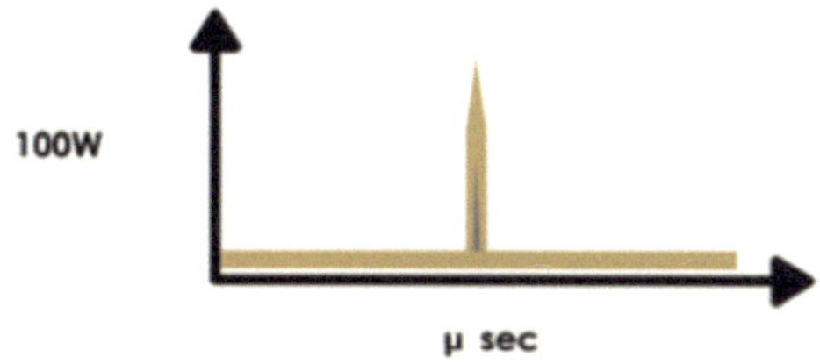

Figure 14.
Super pulsed emission (microsec).

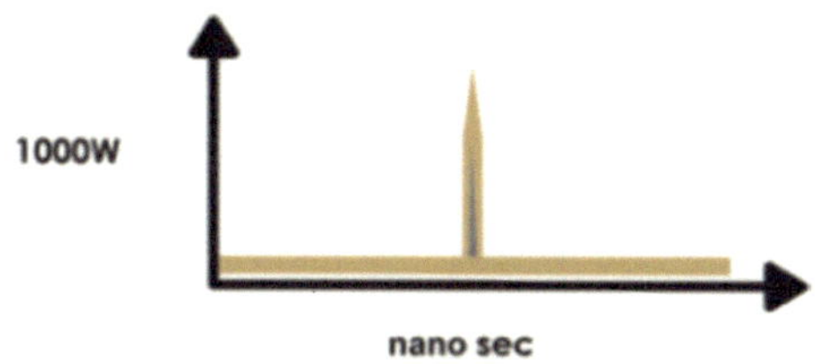

Figure 15.
Ultra pulsed emission (nanosec).

a short wave on target without, however, reaching surrounding tissues and they are mainly used in minimally invasive surgery or photoablation treatments.

4.1 Pulse duration, thermal damage, and energy fluence

Pulse duration is the time over which a laser pulse is delivered, measured in nanoseconds (ns), microseconds (μs), or milliseconds (ms). It should be equal to or shorter than the *thermal relaxation time (TRT)*—the time required for a target to dissipate 50% of absorbed heat—to confine the thermal effects and minimise damage to adjacent tissues. (**Figure 16**).

Thermal damage time (TDT) refers to the time needed to heat the chromophore and surrounding tissue, and is typically longer than the TRT. Effective selective photothermolysis (**Figure 17**) requires pulses enough short to retain heat within the target while reducing collateral damage [25].

Energy fluence (or energy density) is the energy delivered per unit area of the irradiation beam, expressed in J/cm^2 or mJ/cm^2. This, along with beam size, determines penetration depth and energy distribution (**Figure 18**). Beam size refers to the

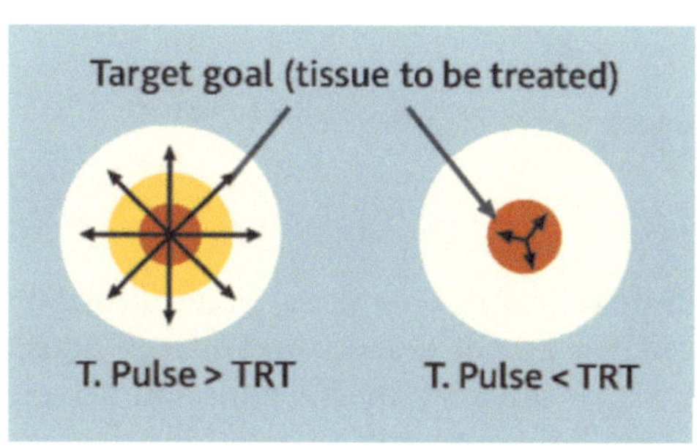

Figure 16.
Pulsed duration.

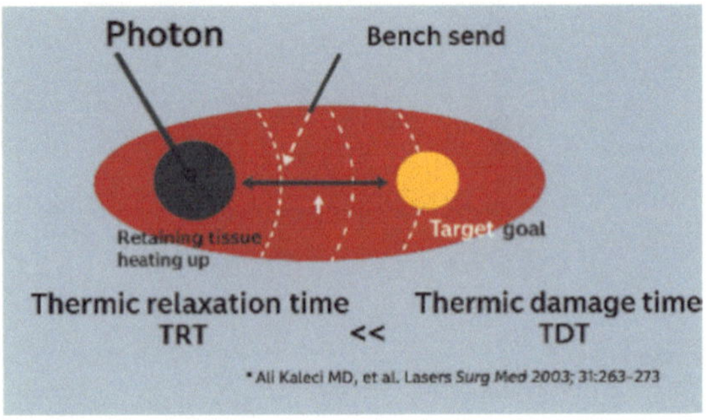

Figure 17.
Selective photothermolysis.

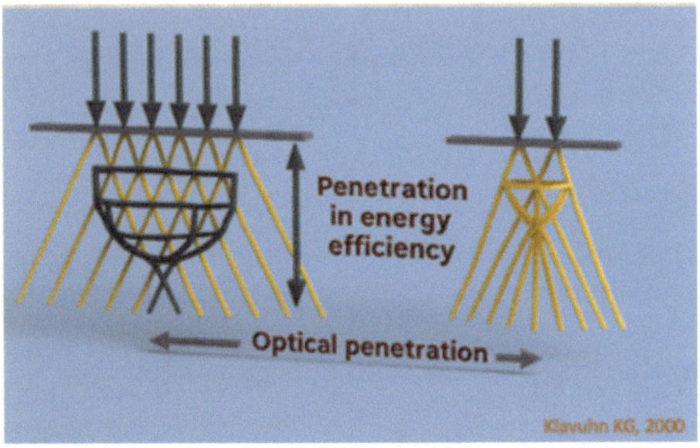

Figure 18.
Energy and optical penetration.

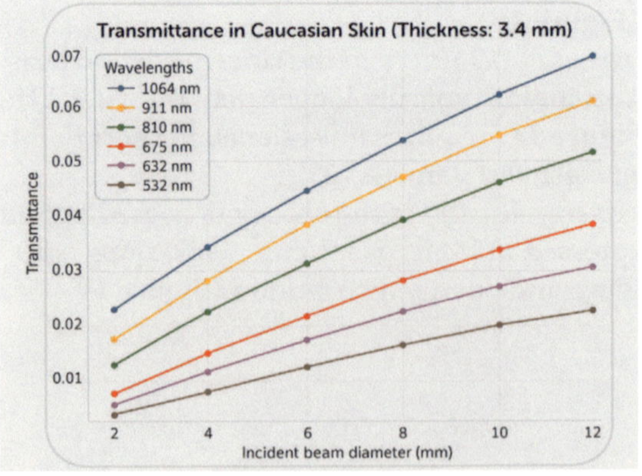

Figure 19.
Skin phototype.

diameter of the photoluminal beam at the point of application and directly affects dosimetric parameters such as power density and energy density.

This characteristic is also influenced by the emission wavelength and the skin phototype (**Figure 19**), with light penetration generally decreasing in more pigmented skin when using small spot sizes. However, for spot sizes greater than 8 mm, differences in penetration depth across different skin types become minimal or negligible.

Photoelectronic (photodisruption or photoacoustic)	• Power density • Photothermocoagulant • Photocytotoxic • Photobiomodular (analgesic, and inflammatory, trophic)
Dosimetric parameters used	• Average power/power and peak power • Energy/pulse/point and total • Power density or irradiance • Energy density or fluence *Spatial overage energy fluence* (SAFE)

Table 3.
Dosimetric parameters.

Respect to the point of application:
• Type of issue • Density of power or irradiance • Energy or energy density • Energy deposited in each punctual or zonal application

Respect to the treated area:
• Total energy deposited per session • SAFE

Table 4.
Parameters for laser's use.

These effects are further modulated by the dispersion of radiation in the tissue and the transmission properties of the medium.

The significance of these data is of great importance since the different effects of Lasers and other photoluminescent systems (IPLS) are related to them, together with the previously described characteristics so that we are able to know what dosimetric parameters we shall apply in a laser treatment (**Tables 3** and **4**).

Those laser's properties are the key casual factors in the reported outcomes related to cheiloplasty scars reduction. [26]

5. Conclusions

Assuming lasers can effectively be used to ablate biological tissues, they can also, and with exactly the opposite effect, help repair internal wounds or holes or scars in relevant organs, so that small defects in the body typically require the suturing of tissues in a surgical process, which as we have seen previously can cause significant and even stigmatising consequences for the patients.

Consequently, in our opinion, when creating and developing, based on our experience this new scar elimination method, the *Teixeira-Pinheiro Technique (2025),* we conclude that laser therapy should be considered the first-line treatment for scars and post-surgical or post-traumatic contractures.

Facial modulation by laser is a therapeutic and/or aesthetic procedure that uses coherent electromagnetic radiation, with controlled parameters such as wavelength, energy density, and pulse duration, to induce the biochemical and biophysical responses in facial tissues.

Effects include stimulation of collagen synthesis, extracellular matrix remodelling, increased mitochondrial activity, and improved microcirculation, promoting skin rejuvenation, tissue regeneration, and local anti-inflammatory control [21, 27–29].

Among the various laser technologies available, the picosecond Nd: YAG laser has proven to be the most appropriate for treating hypertrophic scars following corrective cheiloplasty in patients with cleft lip.

Nevertheless, in some selected cases, particularly those involving significant scar contracture or synechiae requiring excision, the sequential use of a CO_2 laser in combination with a picosecond laser appears to be beneficial [20]. The CO_2 laser may be applied in either *continuous mode*, facilitating tissue *carbonisation*, or in *ultra-pulsed mode*, enabling precise target *volatilisation* [26].

This technique offers clear advantages, including low morbidity, minimal damage to the superficial skin, and both functional and aesthetic improvements. In all of the more than thirty cases treated using this innovative approach, significant clinical improvement was observed and demonstrated.

The primary challenge now lies in determining the optimal parameters for the use of those lasers we have been explaining in these pathologies.

Such optical parameters—specific wavelengths, light energy intensity, individual and treatment regimen durations, and the light source's proximity to the treatment area—must all be carefully considered and accounted for all these factors are essential to achieving favourable therapeutic outcomes.

However, these showed results support the use of picosecond Nd: YAG laser remodelling as a promising and effective technique for the management of post-cheiloplasty scars in patients with cleft lip.

Looking ahead, in the near future it may soon become standard practice to employ nonablative lasers with biomodulation effects—such as the non-surgical laser therapy (NSLT)—immediately after tissue suturing and in the next days following surgery [30].

This proactive approach could help prevent the development of hypertrophic scars after surgical incisions by modulating the early stages of wound healing.

Conversely, as with any novel technique, further studies are required not only to validate its efficacy and safety but also to refine and optimise its use, as laser technology continues to evolve, we anticipate that its role will expand, offering enhanced tools for improving both the aesthetic outcomes and overall, a good health of our patients.

Author details

Heitor Teixeira[1]*, Ruben Pinheiro[2] and Carolina Teixeira[1]

1 CPorto Clinic Porto, Portugal

2 Hospital Lusíadas Porto – Grupo FLAP, Portugal

*Address all correspondence to: hteixeira19@gmail.com

References

[1] Mossey PA, Little J, Munger RG, Dixon MJ, Shaw WC. Cleft lip and palate. Lancet. 2009;**374**(9703): 1773-1785

[2] Dixon MJ, Marazita ML, Beaty TH, Murray JC. Cleft lip and palate: Understanding genetic and environmental influences. Nature Reviews Genetics. 2011;**12**(3):167-178

[3] Jiang R, Bush JO, Lidral AC. Development of the upper lip: Morphogenetic and molecular mechanisms. Developmental Dynamics. 2006;**235**(5):1152-1166

[4] Leslie EJ, Marazita ML. Genetics of cleft lip and cleft palate. American Journal of Medical Genetics Part C Seminars in Medical Genetics. 2013; **163C**(4):246-258

[5] Kondo S, Schutte BC, Richardson RJ, et al. Mutations in IRF6 cause Van der Woude and popliteal pterygium syndromes. Nature Genetics. 2002;**32**(2):285-289

[6] Murray JC. Gene/environment causes of cleft lip and/or palate. Clinical Genetics. 2002;**61**(4):248-256

[7] Little J, Cardy A, Munger RG. Tobacco smoking and oral clefts: A meta-analysis. Bulletin of the World Health Organization. 2004;**82**(3):213-218

[8] Berk NW, Marazita ML. Costs of cleft lip and palate: Personal and societal implications. In: Wyszynski DF, editor. Cleft Lip and Palate: From Origin to Treatment. Oxford: Oxford University Press; 2002. pp. 458-467

[9] Hunt O, Burden D, Hepper P, Johnston C. The psychosocial effects of cleft lip and palate: A systematic review. European Journal of Orthodontics. 2005; **27**(3):274-285

[10] Sommerlad BC, Li S. Sommerlad's technique of unilateral cleft lip repair. In: Shi B, Sommerlad BC, editors. Cleft Lip and Palate Primary Repair. Berlin, Heidelberg: Springer; 2013

[11] Cao X, Wu X, Zhang Y, Qian X, Sun W, Zhao Y. Emerging biomedical technologies for scarless wound healing. Bioactive Materials. 2024;**42**:449-477. ISSN 2452-199X. DOI: 10.1016/j.bioactmat.2024.09.001

[12] Mustoe TA, Cooter RD, Gold MH, et al. International clinical recommendations on scar management. Plastic and Reconstructive Surgery. 2002;**110**(2):560-571

[13] Atiyeh BS. Nonsurgical management of hypertrophic scars: Evidence-based therapies, standard practices, and emerging methods. Aesthetic Plastic Surgery. 2007;**31**(5):468-492

[14] Cordova LZ, Alexander W, Chong D. The role of scar massage in cleft lip surgery. Cleft Lip and Palate. 2021;**4**(2)

[15] Saki N. Picosecond laser applications in aesthetic dermatology. Journal of Surgical Dermatology. 2017;**2**(T1)

[16] Srivastava A, Kale TP, Pathak R, Garg A, Bichile R. Efficacy of lasers on reduction of post-surgical lip scar in cleft lip and palate patients: A systematic review. British Journal of Oral and Maxillofacial Surgery. 2025;**63**(1):3-11. DOI: 10.1016/j.bjoms.2024.09.010

[17] Manstein D, Herron G, Sink K, Tanner H, Anderson R. Fractional photothermolysis: A new concept for cutaneous remodeling using microscopic

patterns of thermal injury. Lasers in Surgery and Medicine. 2004;**34**:426-438. DOI: 10.1002/lsm.20048

[18] Longo L, Postiglione MG, Marangoni O, Melato M. Two-year follow-up results of copper bromide laser treatment of striae. Journal of Clinical Laser Medicine & Surgery. 2003;**21**(3):157-160. DOI: 10.1089/104454703321895617

[19] Teixeira H. Laser in medicine. In: Laser in Dermatology and Other Areas. 2017. pp. 226-231

[20] DeHoratius DM, Dover JS. Nonablative tissue remodeling and photorejuvenation. Clinics in Dermatology. 2007;**25**(5):474-479. ISSN 073081X. DOI: 10.1016/j.clindermatol.2007.05.006

[21] Avci P, Gupta A, Sadasivam M, Vecchio D, Pam Z, Pam N, et al. Lowlevel laser (light) therapy (LLLT) in skin: Stimulating, healing, restoring. Seminars in Cutaneous Medicine and Surgery. 2013;**32**(1):41-52. DOI: 10.12788/j.sder.0005

[22] Karsai S, Raulin C. Laser and IPL technology in dermatology: Basics, indications, and risks. Journal der Deutschen Dermatologischen Gesellschaft. 2007;**5**(12):1035-1043

[23] Anderson RR, Parrish JA. Selective photothermolysis: Precise microsurgery by selective absorption of pulsed radiation. Science. 1983;**220**(4596):524-527. DOI: 10.1126/science.6836297

[24] El Arabi Y, Hali F, Chiheb S. Laser management and safety in dermatology. Cureus. 2022;**14**(6):e25991. DOI: 10.7759/cureus.25991

[25] Altshuler GB, Anderson RR, Manstein D, Zenzie HH, Smirnov MZ. Extended theory of selective photothermolysis. Lasers in Surgery and Medicine. 2001;**29**(5):416-432. DOI: 10.1002/lsm.1136

[26] Kroma-Szal A, Pawlaczyk M, Gornowicz-Porowska J. Medical applications of picosecond lasers for removal of non-tattoo skin lesions—A comprehensive review. Sensors. 2025;**15**(9):4719. DOI: 10.3390/s15094719

[27] Tan J, Zhou J, Huang L, Fu Q, Ao M, Yuan L, et al. Hypertrophic scar improvement by early intervention with ablative fractional carbon dioxide laser treatment. Lasers in Surgery and Medicine. 2021;**53**(4):450-457. DOI: 10.1002/lsm.23301. Epub 2020 Jul 16

[28] Zimber MP, Mansbridge JN, Taylor M, Stockton T, Hubka M, Baumgartner M, et al. Human cell-conditioned media produced under embryonic-like conditions result in improved healing time after laser resurfacing. Aesthetic Plastic Surgery. 2012;**36**(2):431-437. DOI: 10.1007/s00266-011-9787-8. Epub 2011 Jul 7

[29] Alster TS, Lupton JR. Lasers in dermatology. An overview of types and indications. American Journal of Clinical Dermatology. 2001;**2**(5):291-303. DOI: 10.2165/00128071-200102050-00004

[30] Freitas LF, Hamblin MR. Proposed mechanisms of photobiomodulation or low-level light therapy. IEEE Journal of Selected Topics in Quantum Electronics. 2016;**22**(3):7000417

Chapter 9

Multiphysics Modeling of Pulsed Nd-YAG Laser Welding Process for Thermal Field and Residual Stress Prediction

Mouad Bensada and Abdellah Laazizi

Abstract

This chapter presents multiphysics modeling approach of the laser welding process, with a focus on Nd-YAG pulsed laser welding applied to X30Cr13 stainless steel. The goal is to simulate the temperature field, predict the weld pool geometry and residual stress. The proposed model contributes to improving the reliability and performance of welded components in industrial applications. The interaction between the laser beam and the material is modeled through heat transfer equations with appropriate boundary conditions, incorporating material properties and laser as energy source. The numerical simulations are conducted using finite element analysis and are compared to experimental data. The results provide insights into temperature distribution, melt pool dynamics, and residual stress, offering a predictive tool for optimizing welding parameters and determining welding defects.

Keywords: Nd-YAG laser welding, multiphysics modeling, heat transfer, stainless steel, numerical simulation, residual stress

1. Introduction

Laser welding is a high-precision manufacturing technique widely used in industries such as automotive, aerospace, and medical devices. Its ability to deliver localized heat input with minimal distortion makes it ideal for joining high-performance materials. Among the different types of lasers, the Nd-YAG pulsed laser is known for its ability to perform controlled and deep penetration welding with low thermal impact on surrounding zones [1]. However, the localized and rapid heating can generate significant thermal gradients, leading to residual stress and potential defects. Understanding and predicting these effects through numerical modeling is crucial for quality assurance and process optimization [2].

Despite its advantages, the localized and rapid heating-cooling cycles characteristic of pulsed laser welding generate steep thermal gradients within the material. These gradients lead to the development of thermal stresses, which may exceed the elastic limit of the material and initiate plastic deformation or even fracture.

IntechOpen

To address these challenges, this chapter presents a detailed multiphysics modeling framework of the Nd-YAG pulsed laser welding process. The model couples transient heat transfer with realistic laser beam energy input and considers temperature-dependent material behavior [3]. The simulations are applied to X30Cr13 martensitic stainless steel and aim to accurately predict thermal fields, melt pool geometry and residual stress distribution. The numerical results are validated through experimental data, and the developed model that provides a predictive tool for process optimization and defect prevention in high-performance welding applications.

2. Overview of welding processes

Welding is a fundamental joining technique widely used in manufacturing and construction sectors. It involves the coalescence of materials—typically metals—through localized heating, often accompanied by the application of pressure and/or filler materials. Several welding methods have been developed to address the requirements of different applications, materials, and operating conditions. Among these, TIG (Tungsten Inert Gas), MIG/MAG (Metal Inert/Active Gas), and laser welding are prominent for their performance and adaptability.

TIG welding, also known as Gas Tungsten Arc Welding (GTAW), employs a non-consumable tungsten electrode combined with an inert shielding gas such as argon or helium to prevent oxidation. This method produces high-quality and precise welds with minimal contamination, making it particularly suitable for critical applications in sectors like aerospace and nuclear industries, although it remains relatively slow and requires a high level of operator skill [4]. In contrast, MIG/MAG welding (Gas Metal Arc Welding – GMAW) uses a consumable wire electrode that is continuously fed through a welding gun together with a shielding gas. MIG is typically applied with inert gases for non-ferrous metals, while MAG relies on active gases for steel welding. This process is faster than TIG and can be used in multiple positions, though it is more sensitive to environmental conditions and prone to spatter formation [5]. More recently, laser beam welding has emerged as a highly precise and efficient joining method, offering deep penetration, high welding speeds, and minimal heat-affected zones (HAZ). The concentrated energy input results in a narrow fusion zone that reduces thermal distortion, while Nd-YAG pulsed lasers in particular provide accurate thermal control and low energy consumption, making them well-suited for advanced manufacturing applications [6].

During welding, the material undergoes complex thermal and mechanical cycles that give rise to several key physical phenomena. Rapid heating leads to melting of the base material and the formation of a weld pool, followed by solidification upon cooling, which strongly influences the resulting microstructure and mechanical properties. Surrounding regions, known as heat-affected zones (HAZ), do not melt but undergo microstructural changes due to elevated temperatures, often causing localized variations in hardness and toughness. As the weld cools, uneven thermal contraction generates residual stresses that can reduce fatigue strength and compromise dimensional stability. Moreover, defects may occur during solidification, the most common being porosity from entrapped gas bubbles, lack of fusion between the weld and base material, geometric imperfections such as irregular bead shapes or excessive undercut, and cracks or inclusions resulting from foreign particles like tungsten or slag [6]. Understanding the origins of these defects and the associated thermal cycles is essential for developing predictive numerical models, which in turn support process optimization and quality control while reducing reliance on costly trial-and-error experimentation.

3. Multiphysics modeling of the Nd-YAG welding process

This section describes the modeling methodology:

Heat source modeling: The laser is modeled as a Gaussian or surface heat flux with spatial and temporal distribution.

Material properties: Temperature-dependent thermal conductivity, specific heat, and density are incorporated.

Governing equation (2D): The transient heat conduction equation (Eq. 1) [7]:

$$\rho cp\frac{\partial H}{\partial t}=\frac{\partial}{\partial x}\left(k\frac{\partial H}{\partial x}\right)+\frac{\partial}{\partial z}\left(k\frac{\partial H}{\partial z}\right) \tag{1}$$

The enthalpy defined as:
$H=\int_{T0}^{T}Cpdt+Lm\,f\,(T)\,and\;\;Cp = Cp_0 + Lm\,\frac{df}{dT}$ is an equivalent specific heat [7] (Eq. 2):

$$f(T)=\begin{cases}1 & if\;T>T_{liq}\\ \dfrac{T-T_{liq}}{T_{liq}-T_{Sol}} & if\;T_{Sol}\le T\le T_{liq}\\ 0 & if\;T<T_{Sol}\end{cases} \tag{2}$$

Where:
Lm : Latent heat of fusion.
f(T): Fraction of liquid as a function of temperature.

3.1 Laser source modeling

Figure 1 shows a Laser impact on photographic paper at 3 J energy, which includes two circular zones (1 and 2). A two-gaussian axisymmetric is modeled showed by the following equation [8]:

$$\begin{aligned}D_p(r)&=D_{p0}\left[P_1exp\left(\frac{r^2}{r_1^{\,2}}\right)+P_2exp\left(\frac{r^2}{r_2^{\,2}}\right)\right]\\ D_{p0}&=\frac{E}{\pi\tau\left(P_1r_1^{\,2}+P_2r_2^{\,2}\right).}\end{aligned} \tag{3}$$

With:
P_1 : Power fraction of distribution n°1;
P_2 : Power fraction of distribution n°2,
D_{p0} : Homogeneous normalization constant at a power density;
r_1 : Radius corresponding to distribution n°1;
r_2 : Radius corresponding to distribution n°2
Modeling Interaction Laser- Material

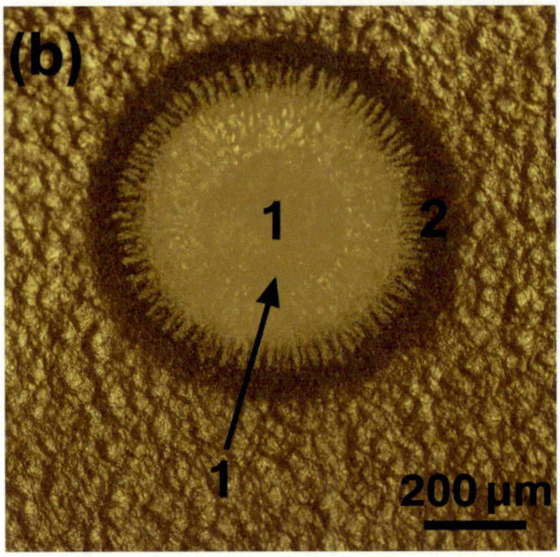

Figure 1.
One pulse laser source impact.

3.2 Used welded material

The used material in this study is Stainless steel X30Cr13, the following table shows the composition (**Table 1**) and the mechanical and thermal characteristic of this material (**Table 2**):

3.3 Resolution of heat transfer equation

The Finite Element Method FEM is used to resolve the 2D model implemented in COMSOL Multiphysics software [9].

Element	Fe	Cr	C	Si	Ni	Mn	P
%	85.92	12.62	0.30	0.63	0.11	0.41	0.01

Table 1.
Material composition.

Melting temperature	~ 1500°C
Density	7.7
Thermal capacity	460 $J.kg^{-1}.K^{-1}$
Thermal conductivity	30 $W.m^{-1}.K^{-1}$
Thermal diffusivity	8.5 10^{-6} $m^2.s^{-1}$
Young modulus	206 GPa
Hardness	4 GPa
Yield strength	850 MPa

Table 2.
Mechanical and thermal characteristics of the material.

3.3.1 Boundary conditions

Top surface:

$$-K\frac{\partial T}{\partial z} = \mathrm{h_{conv}}\left(\mathrm{T}-\mathrm{T_0}\right)$$

$$+ D_{p0}\left[P_1 exp\left(-\left(\frac{x-(x_0+vt)}{r_1}\right)^2\right) + P_2 exp\left(-\left(\frac{x-(x_0+vt)}{r_2}\right)^2\right)\right]\mathrm{X}(t) \quad (4)$$

z = 5 mm: T = T_0

$x = 0$, L: $-K\frac{\partial T}{\partial z} = \mathrm{h_{conv}}\left(\mathrm{T}-\mathrm{T_0}\right)$

With $\mathrm{X}(t)$ is the pulse function:
(See **Figures 2** and **3**)

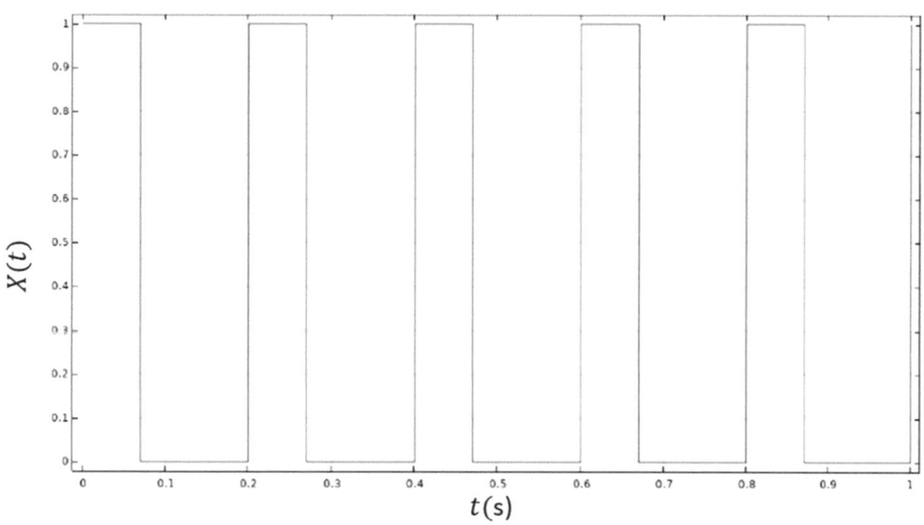

Figure 2.
Pulse of the laser source.

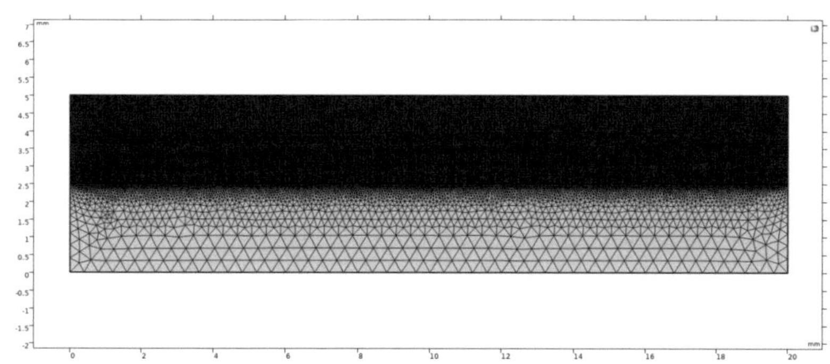

Figure 3.
Meshing of the workpiece.

The computational domain was discretized using a two-dimensional triangular mesh composed of 86,326 elements. To capture the steep temperature gradients in the vicinity of the weld pool, the mesh was locally refined, with the smallest element size reaching 0.04 mm in the fusion zone. Away from this region, a gradual coarsening strategy was adopted to reduce computational overhead while preserving accuracy in the thermal field. Particular attention was paid to maintaining smooth element transitions in order to minimize interpolation errors and ensure stable numerical performance.

The transient heat conduction equation was solved with a finite element formulation using adaptive time stepping. This strategy allowed the solver to automatically reduce the time increment during rapid thermal transients induced by the laser pulse, while employing larger steps during slower cooling phases. Convergence was monitored through temperature residuals and energy balance checks, ensuring that the calculated thermal histories accurately represented the physical process. This combination of spatial refinement and adaptive temporal resolution provided a reliable framework for resolving the melt pool dynamics and predicting heat-affected zone characteristics without excessive computational cost.

3.3.2 Case study

To evaluate the proposed model, a case study involving pulsed Nd-YAG laser welding was simulated under realistic experimental conditions (**Table 3**).

These inputs were used to feed the heat source term in the model and simulate the transient temperature field and residual stress evolution [10].

3.4 Numerical results

3.4.1 Thermal field

The numerical model developed in this study was executed using COMSOL Multiphysics to simulate the temperature distribution and estimate the residual stresses generated during pulsed Nd-YAG laser welding. The objective was to validate the model through experimental data and assess its predictive capability.

The computed temperature fields reveal several characteristic features of pulsed laser welding. At the core of the weld, temperatures exceed the melting point (~1500°C), forming the fusion zone. This central region is surrounded by a heat-affected zone (HAZ), where significant thermal gradients develop despite temperatures remaining below the melting threshold. As the laser advances, a distinct

Energy per pulse power	**$E < 35$ J**
Pulse duration	$0.2 < \tau < 25$ ms
Pulse emission frequency	1 Hz $< F <$ 1000 Hz
Pulse power	$P < 7000$ W
Average emission power	300 W

Table 3.
Experimental conditions.

cooling front is observed, its motion governed by the pulse frequency and the travel speed of the heat source. The overall temperature distribution is strongly dependent on the pulse duration and energy input. In the present case, with a short pulse length of 3.4 ms and a moderate frequency of 145.5 Hz, the model successfully captures the cyclic thermal loading pattern that defines pulsed laser welding.

The melt pool geometry predicted by the model matches well with experimental metallographic observations. Key features include [11]:

- Depth and width within acceptable deviation ranges compared to the weld cross-sections.
- Smooth temperature gradients at the fusion boundaries.
- Slight asymmetry along the welding direction, consistent with heat accumulation effects during successive pulses.

This indicates that the dual-Gaussian heat source and temperature-dependent material properties are effective in capturing the thermal response of the process.

3.4.2 Residual stress

As shown in **Figure 4**, the maximum temperature of the workpiece reaches 2590 K, which generates high-temperature gradients within the welded part and leads to the appearance of thermal stresses that can reach the elastic limits of the materials. The thermal deformation is expressed as (Eq. 5) (**Figure 5**) [6]:

$$\epsilon_{ther} = \alpha\left(T - T_{ref}\right) \tag{5}$$

where α is the thermal expansion coefficient, and T_{ref} is the reference temperature at which the thermal expansion is zero.

Considering the effect of the previous thermomechanical processing of the base metal, the Von Mises stress caused by thermal stress was simulated at different positions of the workpiece. **Figure 6** shows that the thermal stress at the top

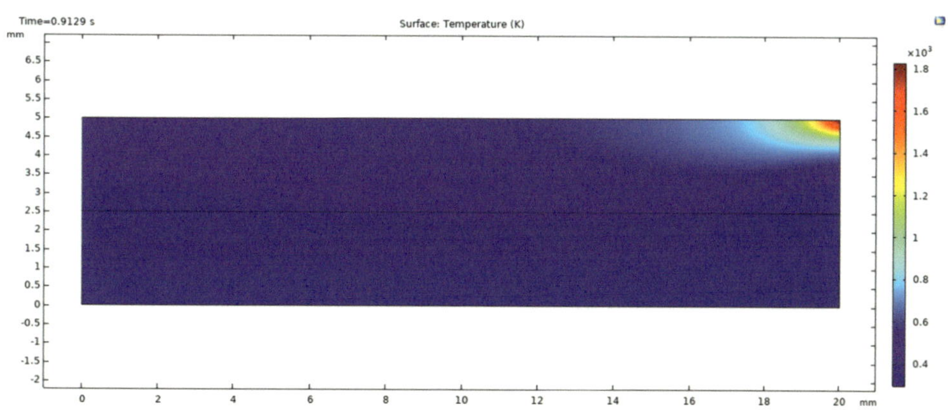

Figure 4.
Temperature field (E = 1.4 J τ = 3.4 ms F = 145.5 Hz V = 21.9 mm/s).

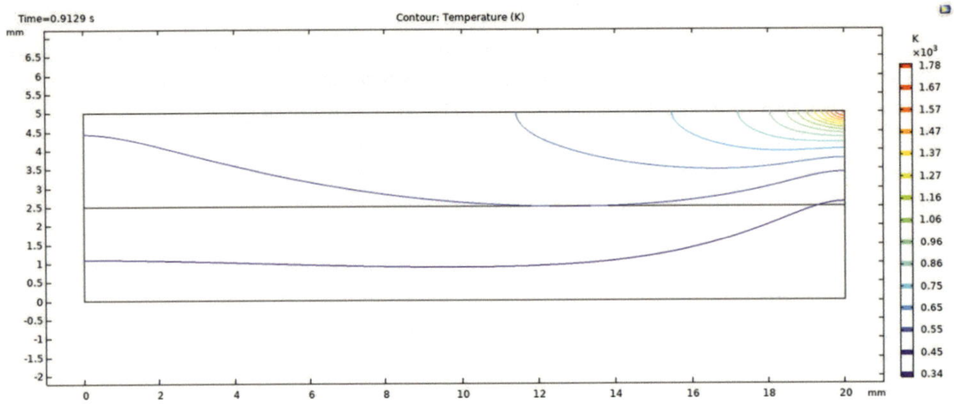

Figure 5.
Isothermals.

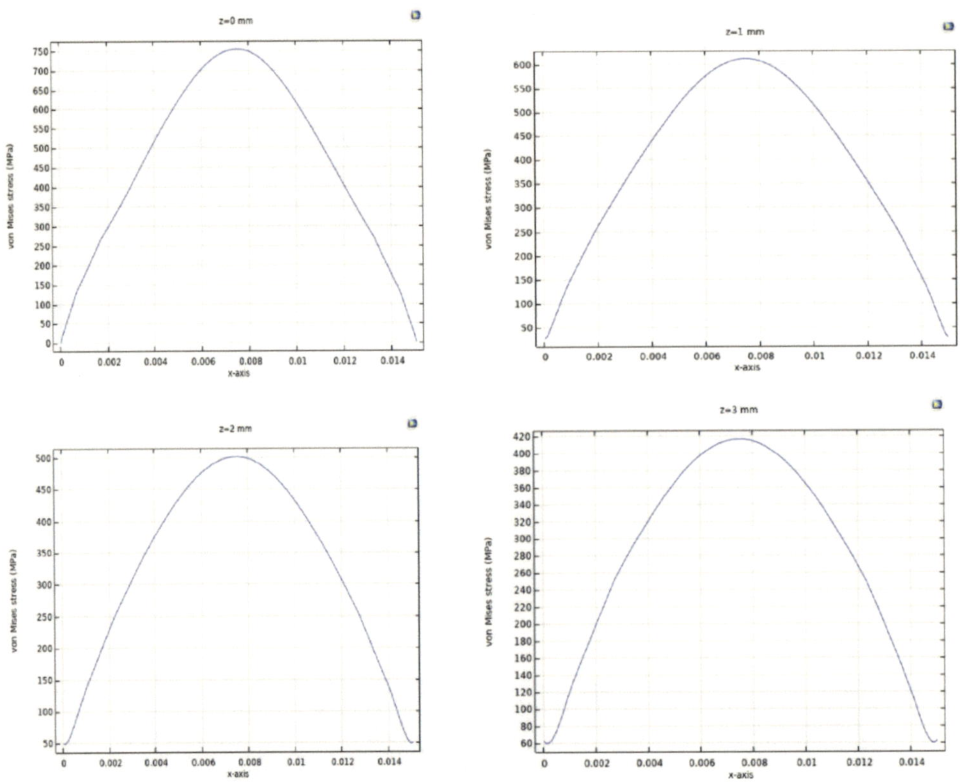

Figure 6.
Evolution of the thermal stress.

surface of the workpiece reaches 750 MPa, which exceeds the yield stress of the material and is near the ultimate tensile strength. This result confirms that the thermal stress induced by the welding process may lead to plastic deformation and may later to material fracture. In this case, controlling the processing parameters is essential for avoiding material fracture and ensuring good weld integrity [12].

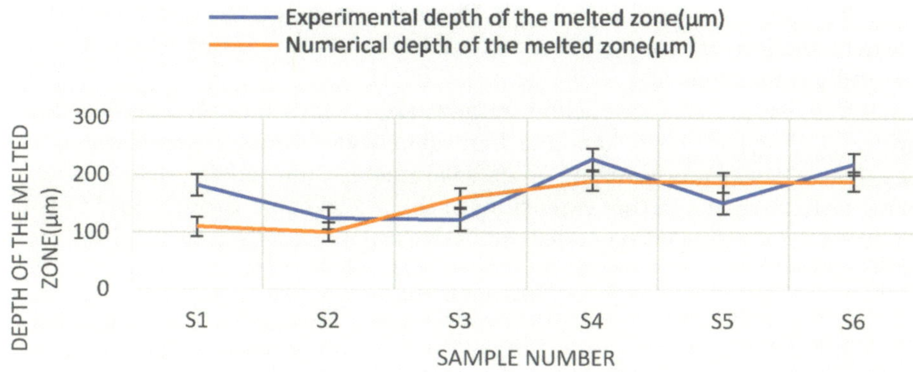

Figure 7.
Comparison numerical vs. experimental results.

3.5 Comparison between simulation and experimental

In order to validate the reliability of the developed model, a systematic comparison was carried out between the numerical predictions and the experimental measurements. **Figure 7** illustrates the correspondence between the calculated and measured penetration depths of the melted zone. Overall, the agreement is satisfactory, with the numerical model reproducing the experimental trends and capturing the general shape and extent of the fusion zone.

The analysis of deviations reveals only a slight underestimation of the melt depth by the simulation. This difference can be attributed mainly to simplifying assumptions introduced during the numerical formulation, particularly the omission of convective flow within the molten pool. Since convection enhances heat transport and promotes deeper penetration in practice, its exclusion leads naturally to a shallower melt pool in the simulation. Despite this limitation, the model successfully reflects the dominant physical phenomena of pulsed laser welding and provides a reliable basis for predicting weld pool geometry and associated thermal fields [13].

4. Conclusion

This chapter presents Multiphysics modeling approach for simulating the Nd-YAG pulsed laser welding process applied to X30Cr13 stainless steel. The developed model integrates a dual-Gaussian heat source, temperature-dependent thermophysical properties, and a transient heat conduction formulation, implemented using the Finite Element Method (FEM) in COMSOL Multiphysics.

Simulation results successfully predict key process outcomes, including the temperature field, melt pool geometry, and residual stress distribution. The validation with experimental data demonstrates good agreement, confirming the robustness and reliability of the model despite simplifications such as neglecting convective flow within the molten pool and reducing the analysis to a 2D axisymmetric domain.

A critical finding of the study is the formation of high thermal stresses within the welded zone, driven by steep temperature gradients. The simulated Von Mises stress values reached up to 750 MPa—surpassing the yield strength of the base material and approaching its ultimate tensile strength. These results suggest a significant risk of plastic deformation, cracking, or even material failure if processing parameters are not carefully controlled.

Therefore, the proposed modeling framework provides not only a predictive tool for understanding the thermal and mechanical behavior during laser welding, but also a valuable basis for optimizing process parameters, minimizing defects, and ensuring weld integrity. Future extensions of the model could include fluid flow in the molten pool offering even greater predictive capabilities for industrial laser welding application.

Author details

Mouad Bensada and Abdellah Laazizi*
Laboratory of Modeling and Multiphysics Engineering (LMIMP), ENSAM-Moulay Ismail University, Meknes, Morocco

*Address all correspondence to: a.laazizi@ensam.umi.ac.ma

References

[1] Biswas S, Pramanik D, Roy N, Biswas R, Kuar AS. An experimental investigation of micro-channels on transparent material using Nd: YAG laser transmission technique. Optics and Laser Technology. 2024;**175**:110671. DOI: 10.1016/J.OPTLASTEC.2024.110671

[2] Stadler M. Etude expérimentale du mouvement hydrodynamique d'un bain métallique et de sa production de vapeurs sur une configuration de soudage TIG. 2016. [Online]. Available from: https://tel.archives-ouvertes.fr/tel-01386578

[3] Garcia-Villarreal A, Sobral H. Design and construction of a pulsed Nd:YAG laser for LIBS applications. Journal of Physics Conference Series. 2015;**582**(1):1-5. DOI: 10.1088/1742-6596/582/1/012012

[4] Obeid O, Alfano G, Bahai H, Jouhara H. Numerical simulation of thermal and residual stress fields induced by lined pipe welding. Thermal Science and Engineering Progress. 2018;**5**:1-14. DOI: 10.1016/J.TSEP.2017.10.005

[5] Santhosh Kumar SNS et al. Microstructure and fracture analysis on MIG welding of nimonic alloys. Materials Today Proceedings. 2023:97-100. DOI: 10.1016/J.MATPR.2023.08.182

[6] Huitron RMP, Vuorinen E. Hot cracking of structural steel during laser welding. IOP Conference Series: Materials Science and Engineering. 2017;**258**(1):0-8. DOI: 10.1088/1757-899X/258/1/012005

[7] Cho WI, Na SJ. Impact of driving forces on molten pool in gas metal arc welding. Welding in the World. 2021;**65**(9):1735-1747. DOI: 10.1007/s40194-021-01138-8

[8] Zhang X, Tang Z, Di S, Wang H, Wu Y, Wang H. Effect of defocus on blue laser spot welding of electrical-steel-laminations. Optics and Laser Technology. 2024;**175**:110716. DOI: 10.1016/J.OPTLASTEC.2024.110716

[9] Anca A, Cardona A, Risso J, Fachinotti VD. Finite element modeling of welding processes. Applied Mathematical Modelling. 2011;**35**(2):688-707. DOI: 10.1016/J.APM.2010.07.026

[10] Avril L. Elaboration de revêtements sur acier inoxydable: Simulation de la fusion par irradiation laser caractérisation structurale, mécanique et tribologique [Phd thesis]. ENSAM Angers; 2003

[11] Duggirala A, Kalvettukaran P, Acherjee B, Mitra S. Numerical simulation of the temperature field, weld profile, and weld pool dynamics in laser welding of aluminium alloy. Optik (Stuttgart). 2021;**247**:167990. DOI: 10.1016/J.IJLEO.2021.167990

[12] Bensada M, Laazizi A, Fri K, Ouzouhou I. Numerical investigation of the effects of driving forces on weld pool convection and thermal stress in a GTAW process. International Journal of Advanced Manufacturing Technology. 2023;**125**(11-12):5185-5196. DOI: 10.1007/s00170-023-10949-6

[13] Bensada M, Laazizi A, Fri K, Fajoui J. Numerical simulation of weld thermal efficiency GTAW process. Lecture Notes in Mechanical Engineering. 2023;**1**:353-362. DOI: 10.1007/978-3-031-23615-0_36

Chapter 10

Ultrashort Pulse Ablation of Metals: Theory and Modelling

Sergio Vela Liñán, Miguel Morales, David Munoz-Martin and Carlos Molpeceres

Abstract

During the last decades, ablation with ultrashort-pulse ($\tau < 10$ ps) lasers has been the subject of extensive research, justified by its wide range of applications. In this chapter, we present theoretical and numerical approaches to model those processes for metals. In USP ablation scenarios, the pulse duration is too short for the lattice to absorb the energy during the pulse, necessitating the use of the two-temperature model (TTM) to describe the system's thermal evolution by separately considering electron and lattice temperatures. Ablation is governed by spallation (a thermo-mechanical process responsible for the ejection of layers of material) near the threshold fluence and by phase explosion (a sudden expulsion of both vapour and equilibrium liquid droplets due to homogeneous nucleation) at higher fluences. Additionally, we present a novel open-access computational model which numerically solves the TTM equations using finite difference methods. The predictions from our model show good agreement with experimental data for the case of aluminium.

Keywords: USP ablation, two-temperature model, spallation, phase explosion, finite difference method

1. Introduction

During the last decades, experimental and numerical research of ultrashort ($\tau < 10ps$) pulsed laser ablation (USP) has gained relevance, justified by its application in thin film fabrication [1], surface modification [2], or micromachining [3]. fs or sub-ps pulses offer advantages over their longer counterparts, such as reduced thermal damage [4] and better ablation efficiencies (defined as volume of material ablated relative to pulse energy [5]).

This chapter will discuss pulsed laser ablation. It will be divided into two main parts. Firstly, we will provide the reader with an overview of the most important theoretical aspects of laser ablation. Subsequently, we will present a code that we have developed to simulate ablation processes in the femtosecond regime.

Figure 1 shows a schematic diagram of pulsed laser ablation. A laser pulse with duration in the range of ns, ps, or fs irradiates a target and heats up the material within a radius from the centre of the laser beam. After melting, if the energy of the laser pulse is high enough, part of the sample will be removed (ablated) by evaporation or

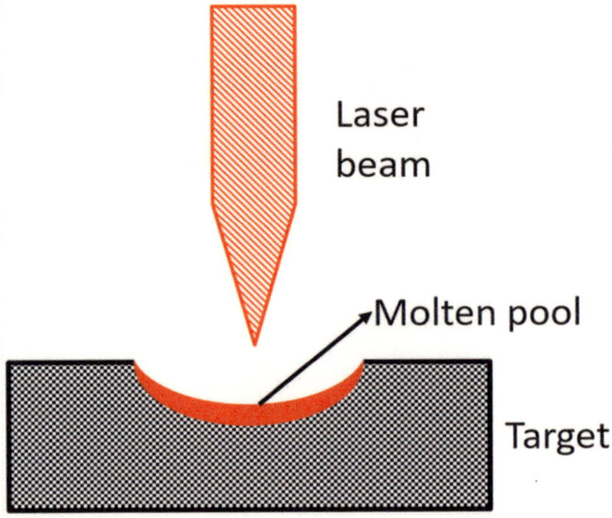

Figure 1.
Schematic diagram of pulsed laser ablation. A laser pulse irradiates a target material and produces melting and subsequent evaporation.

sublimation. Depending on the type of material and on the pulse duration, different mechanisms dominate ablation processes, which will be explained later in this chapter.

2. Theoretical framework

2.1 Basic definitions for a Gaussian beam

If we consider that the laser beam propagates in the $z-$ direction, the general form of the irradiance (sometimes also referred as intensity) distribution is [6]:

$$I_{m,n}(x,y) = I_0 \left[H_m\left(\frac{\sqrt{2}x}{\omega_0}\right)e^{-\frac{x^2}{\omega_0^2}}\right]^2 \left[H_n\left(\frac{\sqrt{2}y}{\omega_0}\right)e^{-\frac{y^2}{\omega_0^2}}\right]^2 \tag{1}$$

Where I_0 is the peak irradiance, ω_0 is the beam waist defined at $1/e^2$, and H_m and H_n are the Hermite polynomials of order m and n, respectively. These ordinals will define the TEM_{mn} modes of the laser (TEM stands for transverse electromagnetic, as both electric and magnetic fields are perpendicular to the direction of propagation). In this chapter, we will always employ the fundamental TEM_{00} mode, which is also called the Gaussian beam:

$$I(r) = I_0 e^{-2\frac{r^2}{\omega_0^2}} \tag{2}$$

From now on, we will employ cylindrical coordinates, as there will be symmetry in the azimuthal angle ϕ. We also need to consider the temporal evolution of the laser pulse. Various approaches exist, but all the lasers that we will use through this work have a Gaussian-type pulse evolution:

$$I(r,t) = I_0 e^{-2\frac{r^2}{\omega_0^2}} e^{-\frac{4\ln(2)(t-t_0)^2}{\tau^2}} \tag{3}$$

Where we have considered that the pulse is centred at a time t_0 and τ is the pulse duration, which is defined as the full-width at half maximum (FWHM). By definition, the power is found by integrating the irradiance distribution over all the surface:

$$\begin{aligned} P(t) &= \int_{Area} I(r,t)dA = \int_0^{2\pi}\int_0^{\infty} I_0 e^{-2\frac{r^2}{\omega_0^2}} e^{-\frac{4\ln(2)(t-t_0)^2}{\tau^2}} r dr d\phi = \frac{\pi\omega_0^2}{2} I_0 e^{-\frac{4\ln(2)(t-t_0)^2}{\tau^2}} \\ &= P_0 e^{-\frac{4\ln(2)(t-t_0)^2}{\tau^2}} \end{aligned} \tag{4}$$

The peak power P_0 is therefore $\pi\omega_0^2/2$. The fluence is defined as the energy per unit area, and thus, it is found by integrating the irradiance over time:

$$F(r) = \int_{-\infty}^{\infty} I(r,t)dt = \int_{-\infty}^{\infty} I_0 e^{-2\frac{r^2}{\omega_0^2}} e^{-\frac{4\ln(2)(t-t_0)^2}{\tau^2}} dt = F_0 e^{-2\frac{r^2}{\omega_0^2}} \tag{5}$$

Where the peak fluence is $F_0 = \tau I_0 \sqrt{\frac{\pi}{4\ln(2)}}$. Finally, the total energy of the laser pulse is calculated by spatial integration of the fluence:

$$E_p = \int_{Area} F(r)dA = \int_0^{2\pi}\int_0^{\infty} F_0 e^{-2\frac{r^2}{\omega_0^2}} r dr d\phi = \frac{F_0 \pi\omega_0^2}{2} \tag{6}$$

Usually, the energy per pulse E_p is the quantity that is directly deduced in a pulsed laser, for example, by measuring the power delivered by the laser over a very long period of time compared to the pulse duration, and then dividing it by the repetition frequency. Therefore, it is useful to rewrite the peak fluence, power, and irradiance as a function of the energy per pulse:

$$\begin{aligned} F_0 &= \frac{2}{\pi\omega_0^2} E_p \\ P_0 &= \sqrt{\frac{\ln(2)}{\pi}} \frac{2}{\tau} E_p \\ I_0 &= \sqrt{\frac{4\ln(2)}{\pi}} \frac{2}{\pi\tau\omega_0^2} E_p \end{aligned} \tag{7}$$

2.2 Heat transfer model in ns regime

If the pulse duration of the laser source is on the order of nanoseconds or longer, the temperature evolution of the material can be described by the classical Fourier heat conduction equation [7]:

$$\rho C_p \frac{\partial T}{\partial t} = \vec{\nabla}\left(k\vec{\nabla}T\right) + Q(r,z,t) \tag{8}$$

Where ρ is the density of the material, C_p its heat capacity at constant pressure, and k is its thermal conductivity. The laser source term $Q(r,z,t)$ should consider the

Gaussian irradiance distribution (3) described in the previous subsection, as well as the reflectivity and absorption of the material. If the later can be described by the Beer–Lambert extinction law (usually the case for metals), this term will be:

$$Q(r,z,t) = I_0(1-R)\alpha e^{-\alpha z} e^{-2\frac{r^2}{\omega_0^2}} e^{-\frac{4\ln(2)(t-t_0)^2}{\tau^2}} \quad (9)$$

Where R is the reflectivity, and α is the absorption coefficient of the material, the rest of variables were described in (3). It should also be noted that, in order to obtain a unique solution, the partial differential eq. (8) requires initial and boundary conditions. The former will be simply the temperature distribution of the piece before laser irradiation, and in single pulse scenarios, all the piece will be initially at room temperature. Regarding the boundary conditions, if the dimensions of the piece are much bigger than the beam radius ω_0 (which is usually the case), it is reasonable to assume that there is no heat flux at the outer edges of the piece:

$$\left(k\vec{\nabla}T\right) \cdot \vec{n} = 0 \quad (10)$$

Where $\vec{n}$ is the normal vector to the surface at the outer edges.

2.3 Heat transfer in fs regime: The two-temperature model

In USP laser ablation, the pulse duration is significantly shorter than the times required for the electron-lattice heat exchange. This means that the classical Fourier heat conduction eq. (8) can no longer be used. When a material is irradiated by an ultrashort pulse, three physical processes will take place. Firstly, free electrons will absorb the energy from the beam during the pulse. This causes the electron temperature to reach very high values, typically in the order of tens of thousands of Kelvin. In the particular case of metals, because their electronic heat capacity is low, this temperature increase will occur without delay with respect to the laser pulse. During this stage, the lattice will remain cold. Then, energy will be diffused to the lattice subsystem, due to electron-phonon collisions, which means that the ion temperature will increase until the electron and lattice subsystems reach thermal equilibrium, the time when this occurs is called the thermalisation time, for metals is usually on the order of tens of picoseconds. And finally, ion-ion collisions result to heat diffusion in the lattice and subsequent cooling.

To model the process described above, the two-temperature model (TTM) is widely used. It was discovered by Kaganov *et al.* [8], and first developed to model USP laser heating by Anisimov *et al.* [9], its governing equations are as follows:

$$\begin{aligned} C_e\frac{\partial T_e}{\partial t} &= \vec{\nabla}\cdot\left(k_e\vec{\nabla}T_e\right) - g(T_e - T_l) + Q(r,z,t) \\ C_l\frac{\partial T_l}{\partial t} &= g(T_e - T_l) \end{aligned} \quad (11)$$

Where T_e and T_l are the electron and lattice temperatures, C_e and C_l are the electron and lattice heat capacities, respectively, k_e is the electron thermal conductivity, g is the electron-phonon coupling factor, and $Q(r,z,t)$ is the laser source term, described already in eq. (9). For many metals, the lattice thermal conductivity k_l is too small compared to the electron thermal conductivity, and it is usually ignored (e.g. for copper k_e at room temperature is 388 $W/(m \cdot K)$ [10] and k_l is around 5 $W/(m \cdot K)$ [11]). In principle, all the four thermal parameters of the model can be temperature-dependent.

The transition from the classical Fourier heat conduction eq. (8) to the TTM occurs when the pulse duration approaches the electron-phonon equilibration time, which can be approximated as $\tau_{ep} = C_e/g$ [4]. τ_{ep} is usually in the range of ps; for example, it situates around 20 ps for gold [12] and around 1 ps for stainless steel [13]. It should also be noted that the standard TTM is only valid if the electron relaxation time τ_e of the material is significantly shorter than the pulse duration [14]. If that is not the case, the two-temperature model needs to be improved, to account for this Qiu and Tien have derived the hyperbolic two-step model (HTS) [15], which adds an extra term to the electron heat conduction equation. Nevertheless, all lasers modelled in this chapter will have significantly longer pulse durations than the electron relaxation time, and therefore, the standard TTM will be enough for our purposes.

2.4 Absorption of laser pulses in metals

As it has been seen in Eq. (9), linear absorption can usually be assumed in metals, which leads to the pulse energy absorption being described by the Lambert-Beer law:

$$Q(r,z,t) = \alpha e^{-\alpha z}(1-R)I_{LS}(r,t) \tag{12}$$

Where α is the absorption coefficient, R is the reflectivity of the material, and I_{LS} is the irradiance distribution produced by the laser source, Eq. (3). To find the reflectivity and the absorption coefficient, we first need to model the relative electric permittivity of the material. The Drude model can be used for this purpose [16]:

$$\epsilon = \epsilon_\infty - \frac{\omega_D^2}{\omega^2 + i\gamma\omega} \tag{13}$$

Where ϵ_∞ is the relative electric permittivity of the unperturbed material, ω_D is the plasma frequency, ω is the angular frequency of the laser, that is $\omega = 2\pi c/\lambda$, and γ is the inverse of the electron relaxation time. To better model the optical properties of the material, several approaches exist. One of them is the Drude + critical points (CP) model [17], which adds one or more additional frequency-dependent terms to eq. (13). On the other hand, several studies have assumed a constant γ; however, better models are available to consider the dependency of the electron collision frequency with temperature or density [18, 19], and this parameter has also been recently measured experimentally during a laser pulse [20].

Once we have found the relative electrical permittivity, the complex refractive index of the metal is given by:

$$\hat{n} = \sqrt{\epsilon} = n + ik \tag{14}$$

n is the usual real refractive index, and k is the extinction coefficient. And finally, the reflectivity and absorption coefficient are given by

$$R = \left|\frac{\hat{n}-1}{\hat{n}+1}\right|^2 \qquad \alpha = \frac{2k\omega}{c} \tag{15}$$

It should also be mentioned that, in some metals, interband absorption mechanisms may also be significant. For example, in the case of aluminium, this phenomenon leads to an increased absorption at certain laser frequencies [19]. To account for this, an additional term can be added to the effective penetration depth, resulting in an effective absorption coefficient given by $(1/\alpha + \lambda_{\text{ball}})^{-1}$, where λ_{ball} accounts for the movement of ballistic electrons [21]. This approach has been used in TTM simulations, showing good agreement with the experimental depths [22].

2.5 Melting and ablation thresholds

The heat conduction equation (either (8) or (11) depending on the pulse duration) will determine a temperature evolution of the irradiated piece. If at any given time its surface temperature reaches the melting temperature, part of the material will undergo melting. Several models exist to account for the liquid phase of the material [23, 24]. If the irradiance is high enough, the vaporisation temperature can also be reached, which will result in part of the material being ejected [25]. Apart from normal vaporisation and boiling, other physical processes can produce ablation, especially for short pulses and/or high fluences, as it will be explained in the next subsection.

This separation between heating, melting, and vaporisation can also be analysed by considering the energy distribution of the beam, as shown in **Figure 2**. Each material will present a characteristic fluence threshold for melting, F_m, and a higher one for vaporisation, F_v. The radial fluence distribution of the laser beam is given by Eq. (5), if the peak fluence F_0 is below F_m (**Figure 2 (a)**), the piece will only heat up, and no phase change will occur. However, if $F_0 > F_m$, a region of the material will melt, which has been marked in light grey in **Figure 2 (b)**. $F_0 > F_v$ case has been shown in **Figure 2 (c)**, and an ablated region (darker grey) will be present. Notice that for this case the molten region (light grey) still exists at the edges of the beam, the extent of which it is normally desirable to reduce for precision manufacturing. This can be achieved by reducing the beam radius ω_0, which will result in a sharper decrease of the fluence close to the threshold for vaporisation F_v.

2.6 Physical mechanisms involved in ablation

Figure 3 schematically shows the dominant physical processes that the target material will undergo as a function of pulse duration and absorbed fluence. Below the melting

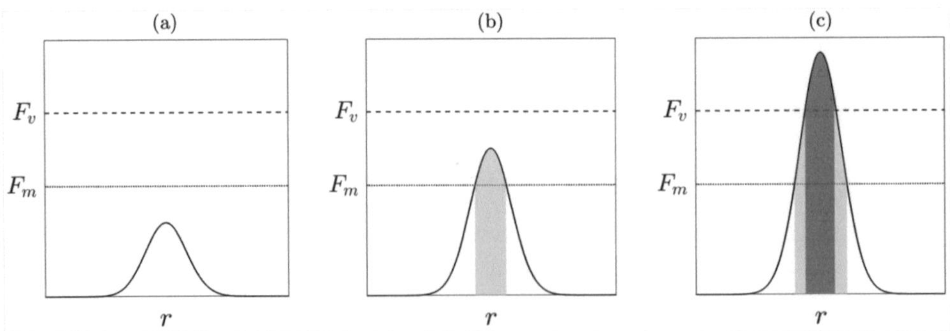

Figure 2.
Radial fluence distribution for three different scenarios. (a) Peak fluence below melting threshold, no phase change will take place. (b) $F_m < F_0 < F_v$, part of the material will undergo melting (light grey region). (c) Peak fluence above vaporisation threshold, an ablated region (dark grey) will coexist with a molten region (light grey).

threshold, no phase change will take place, and the absorbed heat will be dissipated. The melting threshold fluence decreases for shorter pulse durations, as the time scales will reduce the effect of heat dissipation, and therefore, energy will be more efficiently absorbed. For long pulses in the order of microseconds or longer, if the fluence is high, enough ablation will take place *via* normal vaporisation and/or usual boiling. However, if we decrease the pulse duration, there will be no enough time for the conventional phase changes to take place. Under those conditions, phase explosion will be the dominant ablation process at high fluences with spallation appearing at lower fluences only for ultrashort pulses. Those two phenomena will be explained in more detail in the next subsection. At very high fluences, high-temperature plasmas can be generated, a topic that has been the subject of intensive research in the last decades [27, 28], due to its importance for exploring the fundamental physics of relativistic particle beams [29] as well as its usability in recreating extreme stellar phenomena [30].

The actual quantitative values at which the boundaries described above situate depend strongly on the optical and thermal properties of the material, with the melting and ablation thresholds for metals usually being lower than for dielectrics [26]. As the irradiance of the pulse follows a (usually Gaussian) temporal distribution, some of the physical processes described above can occur sequentially. For example, if a metal is irradiated with a sub-ps pulse (vertical dashed arrow on the left in **Figure 3**), at the beginning the energy of the pulse will produce heating and subsequent melting. Then, the part of the material will be expelled *via* spallation. If the fluence is high enough, phase explosion can also occur after an initial spallation layer has been removed. On the other hand, for a ns pulse (central dashed arrow in **Figure 3**), spallation will be absent. Normal vaporisation or boiling will follow the initial heating and melting. If the peak fluence is high enough, phase explosion can also take place at the end of the pulse. For much longer pulse durations near to the continuous wave regime (dashed arrow on the right in **Figure 3**), higher fluences are required to ablate material *via* normal vaporisation due to the heat diffusion, making the process less efficient.

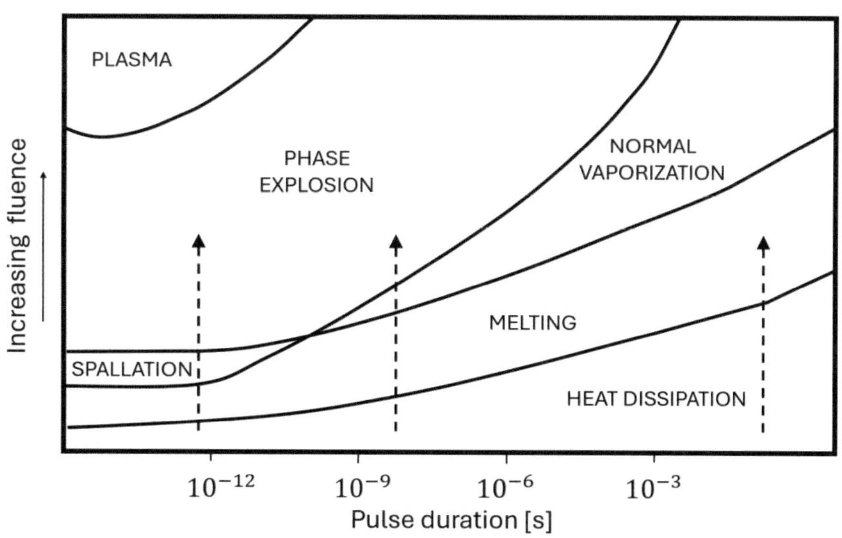

Figure 3.
Responsible mechanisms for ablation as a function of the pulse duration and the absorbed laser fluence. The vertical dashed arrows show the sequence of changes that a sample target material will suffer as the absorbed intensity increases over time, depending on the pulse duration. Figure sketched from the qualitative information presented in Ref. [26].

2.6.1 Spallation and phase explosion

As we have seen in **Figure 3**, for ultrashort pulses two main mechanisms dominate ablation: spallation and phase explosion. Spallation dominates for ps and sub-ps pulses at intensities slightly above the ablation threshold, while phase explosion requires higher intensities. Those phenomena will be now explained in more detail.

We begin with spallation. If the laser irradiates a volume of the target, each material will present a characteristic time τ_s, which is defined as the time required to initiate the motion of atoms within the heated volume. Let us call the pulse duration τ_p and the characteristic time for the energy transfer from electrons to lattice τ_{e-ph}, if the following condition is fulfilled

$$max\left\{\tau_p, \tau_{e-ph}\right\} \leq \tau_s \tag{16}$$

The heating and melting of the metal will take place at constant volume, as there will be no time for the material to expand. The consequence will be a high compressive stress, and therefore, the condition (16) is referred to as the stress confinement condition. (Note that in femtosecond regime the left hand side of (16) will take the value of τ_{e-ph}, which normally is in the order of tens of picoseconds). The compressive stress will interact with the free surface of the target, creating a tensile stress. Above a certain threshold fluence, this tensile stress will be enough to cause mechanical fracture of the target if the surface remains in solid state, or sputtering of liquid droplets; otherwise, in either case it will result in the ejection of a layer of material via spallation [31].

If we increase the energy of the pulse, phase explosion will appear. After melting, the metal can be heated well over the boiling point, because the time scale is too short for heterogeneous bubble nucleation (the phenomenon which produces normal boiling) to take place. This state is referred to as superheating. However, when the lattice temperature reaches approximately 90% of the critical temperature of the material, the kinetic energy of atoms or molecules is so high that they form tiny clusters or "nuclei" of vapour. These clusters form homogeneously throughout all the liquid volume, as opposed to just at specific points such as impurities or surfaces, which would be the case for heterogeneous nucleation. That is why this process is called homogeneous nucleation. Once these vapour bubbles are formed, they grow rapidly due to the high energy input, which will results in a sudden explosion, that ultimately results in the ejection of liquid droplets from the surface of the material [32–34].

The stress confinement condition (16) limits the pulse duration of the laser to a few ps at most in order for spallation to be present, and indeed simulations based on molecular dynamics (MD) have found that this phenomenon is already absent for pulse durations of 50 ps [35]. On the other hand, phase explosion can be observed even for pulse durations in the nanosecond or slightly longer range, as shown in **Figure 3**. Nevertheless, ablation by this mechanism is more efficient for pulse durations in the short ps or fs range.

3. Numerical modelling of USP ablation

We have simulated femtosecond laser ablation of metals using a MATLAB code, which uses finite difference methods (FDM) to numerically solve the TTM eq. (11). Most of the results presented in this section have been published in the following paper: [36].

3.1 Finite difference methods

In this subsection, we will briefly introduce the finite difference scheme. A finite difference quotient of a differentiable function f around the point x is used to approximate $f'(x)$, and it is an expression of the form:

$$\frac{f(x+b)-f(x-a)}{a+b} \tag{17}$$

where a and b are small, non-negative real numbers verifying $a+b \neq 0$. The derivative of f can be approximated by forward, backward, or central difference, depending on the values of a and b:

$$\begin{aligned}
&\text{Forward difference} \Rightarrow b = h \text{ and } a = 0 \Rightarrow f'(x) \approx \frac{f(x+h)-f(x)}{h} \\
&\text{Backward difference} \Rightarrow b = 0 \text{ and } a = h \Rightarrow f'(x) \approx \frac{f(x)-f(x-h)}{h} \\
&\text{Central difference} \Rightarrow b = h \text{ and } a = h \Rightarrow f'(x) \approx \frac{f(x+h)-f(x-h)}{2h}
\end{aligned} \tag{18}$$

Analogously, the second derivatives can be approximated by the following expressions

$$\begin{aligned}
&\text{Forward} \Rightarrow f''(x) \approx \frac{f'(x+h)-f'(x)}{h} \approx \frac{\frac{f(x+2h)-f(x+h)}{h} - \frac{f(x+h)-f(x)}{h}}{h} \\
&= \frac{f(x+2h)-2f(x+h)+f(x)}{h^2} \\
&\text{Backward} \Rightarrow f''(x) \approx \frac{f'(x)-f'(x-h)}{h} \approx \frac{\frac{f(x)-f(x-h)}{h} - \frac{f(x-h)-f(x-2h)}{h}}{h} \\
&= \frac{f(x)-2f(x-h)+f(x-2h)}{h^2} \\
&\text{Central} \Rightarrow f''(x) \approx \frac{f'(x+h)-f'(x-h)}{2h} \approx \frac{\frac{f(x+2h)-f(x)}{2h} - \frac{f(x)-f(x-2h)}{2h}}{h} \\
&= \frac{f(x+2h)-2f(x)+f(x-2h)}{4h^2}
\end{aligned} \tag{19}$$

The errors in the above approximations can be derived from Taylor's theorem. The two-point first derivatives approximated by both backward and forward differences are accurate to $O(h)$, and so are their three-point second derivatives approximations. On the other hand, the central differences presented above are accurate to $O(h^2)$. To obtain higher accuracies, more points on the differences need to be considered. Reference [37] provides the degree of accuracy of central, forward, and backward finite differences depending on the number of points used.

3.2 Numerical resolution of the TTM equations

In order to simulate an USP, the TTM eq. (11) need to be numerically solved. We have created a MATLAB code to perform this task, and in this subsection, the resolution method will be briefly explained. We will employ cylindrical coordinates, and the Laplacian in this coordinate system of a smooth function $f(r, \phi, z)$ is given by:

$$\nabla^2 f = \frac{1}{r}\frac{\partial}{\partial r}\left(r\frac{\partial f}{\partial r}\right) + \frac{1}{r^2}\frac{\partial^2}{\partial \phi^2} + \frac{\partial^2 f}{\partial z^2} \tag{20}$$

The laser beam has symmetry in the azimuthal angle ϕ, which means that there will be no dependency of the temperature field in this coordinate. Therefore, the TTM equations will read:

$$\begin{aligned} C_e\frac{\partial T_e}{\partial t} &= \frac{k_e}{r}\frac{\partial}{\partial r}\left(r\frac{\partial T_e}{\partial r}\right) + k_e\frac{\partial^2 T_e}{\partial z^2} - g(T_e - T_l) + Q(r,z,t) \\ C_l\frac{\partial T_l}{\partial t} &= g(T_e - T_l) \end{aligned} \tag{21}$$

As there is a $\propto 1/r$ term in the equations, we have modified our code to explicitly avoid a zero (or too small) entry in the r array. For each electron or lattice temperature fields, we have created a $(1, 2)$ tensor:

$$T_{(e,l)} = T^n_{(e,l)\ \ k,i} \tag{22}$$

i is the corresponding index for the r coordinate, k is for z coordinate, and n stands for time [36]. The time derivative is approximated by a two-point forward difference (accurate to $O(\Delta t)$, see eq. (18)), and the first derivative with respect to r in the Laplacian is approximated by two-point central difference (accurate to $O\left((\Delta r)^2\right)$), and the second derivatives are approximated by five-point central difference (accurate to $O\left((\Delta r, \Delta z)^4\right)$, see **Table 1** of reference [37]), which results in the following expressions:

$$\begin{aligned} \frac{\partial^2 T_e}{\partial r^2} &\approx \frac{-T^n_{e,k,i-2} + 16T^n_{e,k,i-1} - 30T^n_{e,k,i} + 16T^n_{e,k,i+1} - T^n_{e,k,i+2}}{12(\Delta r)^2} \\ \frac{\partial^2 T_e}{\partial z^2} &\approx \frac{-T^n_{e,k-2,i} + 16T^n_{e,k-1,i} - 30T^n_{e,k,i} + 16T^n_{e,k+1,i} - T^n_{e,k+2,i}}{12(\Delta z)^2} \\ \frac{\partial T_e}{\partial r} &\approx \frac{T^n_{e,k,i+1} - T^n_{e,k,i-1}}{2\Delta r} \\ \frac{\partial T_{(e,l)}}{\partial t} &\approx \frac{T^{n+1}_{(e,l),k,i} - T^n_{(e,l),k,i}}{\Delta t} \end{aligned} \tag{23}$$

a_0	**0.5499**	a_3	**−0.0447**	a_6	**−0.0138**
a_1	−0.0905	a_4	−0.0307	a_7	−0.0089
a_2	−0.0632	a_5	−0.0210	a_8	−0.0054

Table 1.
Fourier coefficients to be used in Eq. (37).

Substituting these differences in (21), we arrive to the following recursive equation, which has been implemented in the code:

$$\begin{aligned}
T^{n+1}_{e,\ k,i} = &-\frac{\Gamma\Delta t}{12(\Delta r)^2}T^n_{e,k,i-2}+\left(\frac{4\Gamma\Delta t}{3(\Delta r)^2}-\frac{\Gamma\Delta t}{2r_i\Delta r}\right)T^n_{e,k,i-1}\\
&+\left(1-\frac{5\Gamma\Delta t}{2(\Delta r)^2}-\frac{5\Gamma\Delta t}{2(\Delta z)^2}\right)T^n_{e,k,i}+\left(\frac{4\Gamma\Delta t}{3(\Delta r)^2}+\frac{\Gamma\Delta t}{2r_i\Delta r}\right)T^n_{e,k,i+1}\\
&-\frac{\Gamma\Delta t}{12(\Delta r)^2}T^n_{e,k,i+2}-\frac{\Gamma\Delta t}{12(\Delta z)^2}T^n_{e,k-2,i}+\frac{4\Gamma\Delta t}{3(\Delta z)^2}T^n_{e,k-1,i}+\frac{4\Gamma\Delta t}{3(\Delta z)^2}T^n_{e,k+1,i}\\
&-\frac{\Gamma\Delta t}{12(\Delta z)^2}T^n_{e,k+2,i}-\frac{g\Gamma\Delta t\left(T^n_{e,k,i}-T^n_{l,k,i}\right)}{k_e}+\frac{\Gamma Q(r_i,z_i,t_i)\Delta t}{k_e}\\
T^{n+1}_{l,\ k,i} = &\ T^n_{l,\ k,i}+\frac{g\Delta t\left(T^n_{e,k,i}-T^n_{l,k,i}\right)}{C_l}
\end{aligned} \tag{24}$$

Where $\Gamma = k_e/C_e$ and the heat source term $Q(r_i, z_i, t_i)$ is the one given in Eq. (9).

$$Q(r_i, z_k, t_n) = I_0(1-R)\alpha e^{-\alpha z_k} e^{-2\frac{r_i^2}{\omega_0^2}} e^{-\frac{4\ln(2)(t_n-t_0)^2}{\tau^2}} \tag{25}$$

Note that we are considering that the laser is irradiating at $z = 0$, which in our case corresponds to the bottom of the piece. All thermal (and optical) parameters are, in principle, temperature-dependent and will be updated for each element with the electron and lattice temperatures registered in the previous time step.

3.2.1 Boundary conditions

The boundary conditions in all edges of the piece will be given from Eq. (10):

$$\left(\vec{\nabla} T_e\right)\cdot\vec{n} = \left(\vec{\nabla} T_l\right)\cdot\vec{n} = 0 \tag{26}$$

Our piece is modelled as a cylinder of radius R and height L, and therefore, this translates to:

$$\begin{aligned}
\left.\frac{\partial T}{\partial r}\right|_{r=-R} = \left.\frac{\partial T}{\partial r}\right|_{r=R} = 0\\
\left.\frac{\partial T}{\partial z}\right|_{z=0} = \left.\frac{\partial T}{\partial z}\right|_{z=L} = 0
\end{aligned} \tag{27}$$

3.2.2 Ablation modelling

To model the elimination of material, we will consider a normal mesh velocity, which ablates material from the bottom of the piece. This method is based on energy conservation and has been used previously by other groups [38].

$$v_n(r) = h_0\cdot\left(\frac{T_l(r)-T_a}{\rho L_v}\right) \tag{28}$$

where ρ is the density, and L_v the latent heat of vaporisation. The numerical parameter h_0 has been set in our case at $h_0 = 5\cdot 10^{13}\ \text{m}^4\text{J}/(\text{s}\cdot\text{K})$.

The ablation temperature T_a is not the usual vaporisation temperature of the metal; instead, it is $0.9T_c$, where T_c is the critical thermodynamic temperature of the metal. This is because phase explosion will dominate at this time scales and fluences [32–34], as it was explained in Section 2.6. For the particular case of Al, the critical temperature is $T_c = 6700\ \text{K}$ [39].

If at a time step t_n, the lattice temperature at a coordinate r_i exceeds the ablation temperature, Eq. (28) will determine an ablation velocity, $v_n(t_n, r_i)$. Therefore, the new height of the piece at this r coordinate is calculated by subtracting $v_n(t_n, r_i) \cdot \Delta t$ to the previous height of the piece. Then, Δz is updated by dividing the new height of the piece by the number of z-elements, which is always kept constant during the simulation. The consequence is that the finite difference Δz changes, and it is a function of both time and radial coordinate. After ablation, the temperature fields T_e and T_l are calculated by interpolation from the previous temperature fields [36].

3.3 Thermal and optical parameters of Aluminium

We have used our code to simulate single pulse ablation on aluminium. For this task, the thermal parameters k_e, g, C_e, and C_l from Eq. (21), and the reflectivity and absorption coefficient (R and α in Eq. (25), respectively) needed to be modelled.

3.3.1 Electron heat capacity C_e

For electron temperatures below 50 kK, the electron heat capacity has been modelled by this cubic approximation [40]:

$$C_e(T_e) = \gamma T_e + \beta T_e^3 \tag{29}$$

For Al, $\gamma = 91.05\ \text{J}/(\text{m}^3\text{K}^2)$ and $\beta = -1.47 \cdot 10^{-8}\ \text{J}/(\text{m}^3\text{K}^4)$. If higher electron temperatures are reached, we would need to use the complete solution for the electron heat capacity, which has also been published in ref. [40] (in our simulations, sometimes that was the case for the highest fluences). We have represented both functions in **Figure 4 (a)**.

3.3.2 Electron thermal conductivity k_e

For the electron thermal conductivity, we need to consider both the electron-electron and electron-ion contribution, and the total electron thermal conductivity is then calculated *via* the Matthiessen rule:

$$\frac{1}{k_e} = \frac{1}{k_{ee}} + \frac{1}{k_{ei}} \tag{30}$$

For the electron-electron contribution, we have used the fitting given in the supplementary material of [41]:

$$k_{ee} = \frac{1 - 0.91906t^{\frac{1}{2}} + 0.79564t + 0.13456t^2}{4 \cdot 10^{-4}t} \tag{31}$$

Where $t = 6k_B T_e/\mu_F$, with k_B being the Boltzmann constant, and μ_F is the Fermi energy (for aluminium, 11.7 eV). The electron-ion contribution is calculated using the following expression [42]:

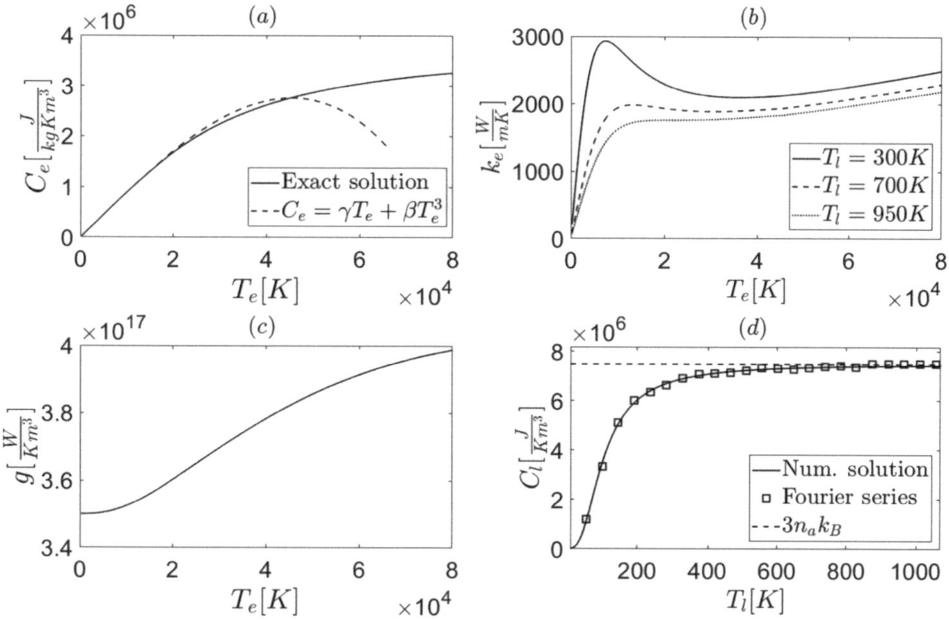

Figure 4.
Two-Temperature model parameters for Al. (a) Electron heat capacity $C_e(T_e)$ both cubic approximation and complete solution from [40] are represented. (b) Electron thermal conductivity for three different lattice temperatures. (c) Electron-phonon coupling factor, model described in [41]. (d) Lattice heat capacity, both solutions obtained from numerically integrating in eq. (36) and Fourier series (37) have been represented.

$$k_{ei} = \frac{1}{3} C_e(T_e) v_s(T_e) \lambda_{ei}(T_l) \tag{32}$$

We have used the same electron heat capacity $C_e(T_e)$ described in the previous subsection. v_s is the mean velocity of s (or p) electrons, given by:

$$v_s(T_e) = \sqrt{v_F^2 + 3\frac{k_B T_e}{m_s}} \tag{33}$$

Where v_F is the Fermi velocity. For Al, m_s is 1.05 times the standard electron mass. Finally, $\lambda_{ei}(T_l)$ is the mean free path between successive electron-ion collisions, which is in general inversely proportional to the lattice temperature [36]. For the sake of simplicity, we have taken the value of k_{ei} at room temperature ($T_{rt} = 298$ K) of $k_{rt} = 237\,\mathrm{W}/(\mathrm{m}\cdot\mathrm{K})$ [43], which allows us to compute the this contribution for any electron temperature:

$$k_{ei} = k_{rt} \frac{C_e(T_e) v_s(T_e) T_{rt}}{C_e(T_{rt}) v_s(T_{rt}) T_l} \tag{34}$$

In **Figure 4(b)**, we have represented the total electron thermal conductivity k_e for three different lattice temperatures. As expected, k_e will be lower for higher lattice temperatures.

3.3.3 Electron-lattice coupling factor g

For this parameter, various models exist. We have also chosen the fitting presented in [41]:

$$g(T_e) = a_0 + a_1 \frac{t_r^{3.0}}{1 + t_r^{0.5} + 0.9t_r^{1.9} + 0.0224t_r^{3.7}} \tag{35}$$

Where the reduced temperature t_r is given by $t_r = (T_e/T_{sc})$; and the scaling temperature is $T_{sc} = 11605K$ (converts K to eV). $a_0 = 3.5 \cdot 10^{17}\ \frac{\mathrm{W}}{\mathrm{m^3K}}$ and $a_1 = 1.0 \cdot 10^{16}\ \frac{\mathrm{W}}{\mathrm{m^3K}}$. This parameter has been represented in **Figure 4(c)**.

3.3.4 Lattice heat capacity C_l

From the Debye theory, this parameter is given by [44]:

$$C_l(T_l) = 9n_a k_B \left(\frac{T_l}{T_D}\right)^3 \int_0^{\frac{T_D}{T_l}} \frac{x^4 e^x}{(e^x - 1)^2} dx \tag{36}$$

Where n_a is the atom density, and T_D the Debye temperature (for aluminium, $n_a = 1.81 \cdot 10^{29}\ \mathrm{m}^{-3}$ and $T_D = 428\ \mathrm{K}$). For lattice temperatures significantly higher than T_D, the lattice heat capacity takes a constant value of $3n_a k_B$. Below this limit, we can either integrate numerically in (36) for each T_l, or we can also use the following Fourier series (the derivation can be found in ref. [36]):

$$C_l(T_l) = \begin{cases} 9n_a k_B \left(\frac{a_0}{2} + \sum_{n=1}^{8} a_n \cos\left(\frac{n\pi T_l}{2T_D}\right)\right) & \text{for } T_l < 2T_D \\ 3n_a k_B & \text{for } T_l \geq 2T_D \end{cases} \tag{37}$$

With the corresponding Fourier coefficients shown in **Table 1**. In **Figure 4(d)**, it can be observed that the constant value approximately works for temperatures above twice the Debye temperature. Both (numerical integration and Fourier series) approaches have been used for this parameter in simulations [36].

3.3.5 Reflectivity and absorption coefficient

To model the optical parameters of aluminium, we will employ the Drude + critical points (CP) model for the electrical permittivity, which was already mentioned in Section 2.4. This model adds extra terms to the Drude model for the complex electric relative permittivity of the material [17]:

$$\epsilon = \epsilon_\infty - \frac{\omega_D^2}{\omega^2 + i\gamma\omega} + \sum_{p=1}^{N} A_p \Omega_p \left(\frac{e^{i\phi_p}}{\Omega_p - \omega - i\Gamma_p} + \frac{e^{-i\phi_p}}{\Omega_p + \omega + i\Gamma_p}\right) \tag{38}$$

The high-frequency limit dielectric constant ϵ_∞ can be taken as one for aluminium [45]. ω is the angular frequency of the laser radiation, that is $\omega = \frac{2\pi c}{\lambda}$ with λ being the wavelength of the laser beam, and ω_D is the plasma frequency (for Al, $2.06 \cdot 10^{16}\ \mathrm{rad/s}$). The critical points parameters have been taken from **Table 2** of ref. [46], with $N = 2$, and are reproduced in **Table 2** of this chapter. γ is the electron collision frequency, which is by definition the inverse of the electron relaxation time, $\gamma = \tau_e^{-1}$, for this parameter we have taken the model presented in the supplementary material of [41], with constant density for our case ($x = 1$). The collision frequency is given by the sum of two contributions, due to electron-electron (ee) and electron-ion (ei) interactions:

	A_p	ϕ_p	Ω_p [rad/s]	Γ_p [rad/s]
$p = 1$	5.2306	− 0.51202	$2.2694 \cdot 10^{15}$	$3.2867 \cdot 10^{14}$
$p = 2$	5.2704	0.42503	$2.4668 \cdot 10^{15}$	$1.7731 \cdot 10^{15}$

Table 2.
Critical points parameters for Al, data taken from [46].

$$\begin{aligned} \gamma &= \gamma_{ee} + \gamma_{ei} \\ \gamma_{ee} &= \frac{\eta_{ee} a_0 t^2 (1 + a_1 t^{p_1})}{1 + a_2 t^{p_2}} \\ \gamma_{ei} &= \gamma_{rts} + \eta_s \gamma_{rts} \frac{T_l}{T_{rt}} \end{aligned} \tag{39}$$

With t defined as in Eq. (31), $t = 6k_B T_e/\mu_F$, and the room temperature has been set at $T_{rt} = 293$ K. The rest of the parameters taken from [41] are shown in **Table 3**.

Once we have the electrical permittivity fully determined, the complex refractive index, reflectivity, and absorption coefficient are given by Eqs. (14) and (15), from Section 2.4.

γ_{rts} [1/s]	η_s	η_{ee}	a_0 [1/s]	a_1	p_1	a_2	p_2
$1.167 \cdot 10^{14}$	0.5175	0.75	$4.9188 \cdot 10^{14}$	1.2	0.75	0.8	2.25

Table 3.
Parameters for Eq. (39) data taken from [41].

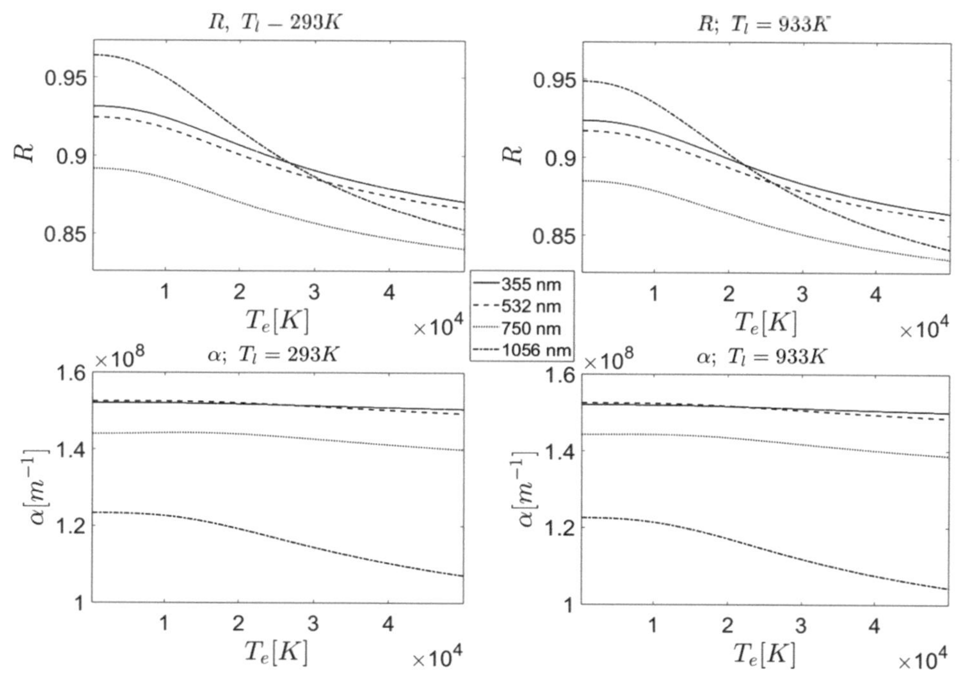

Figure 5.
Reflectivity R and absorption coefficient α at room temperature (left) and at the melting point of Al, which is 933 K.

In **Figure 5**, we have plotted both R and α as a function of electron temperature for both room temperature and a temperature of 933 K. The lasers modelled on this chapter will have their wavelengths in the infrared range, and the $\lambda = 1056$ nm curves shows that there can be a considerable variation of the reflectivity when increasing electron temperature to tens of thousands of Kelvin.

3.4 Results

3.4.1 Experimental procedure

Single pulse ablation experiments were conducted using a diode-pumped solid state laser FemtoLux30 (Ekspla). The pulse duration was variable, and we have performed experiments at 320 fs (value at FWHM). The laser emits at $\lambda = 1030$ nm with a beam radius of approximately 11 μm defined at $1/e^2$. The system is equipped with a galvanometric scanner for beam control (Raylase Superscan-III).

A 6061-T6 sample with an aluminium content over 96 % was used for the experiments. To achieve a flat surface, it was sanded and subsequently polished with polycrystalline diamond suspension, and the grain size was 3 μm. Spots were observed using both a 20× Mitutoyo Plan Apo NIR B Infinity Corrected Objective and a Leica DCM 3D confocal microscope, and an example of a spot observed with the former device is shown in **Figure 6** [36].

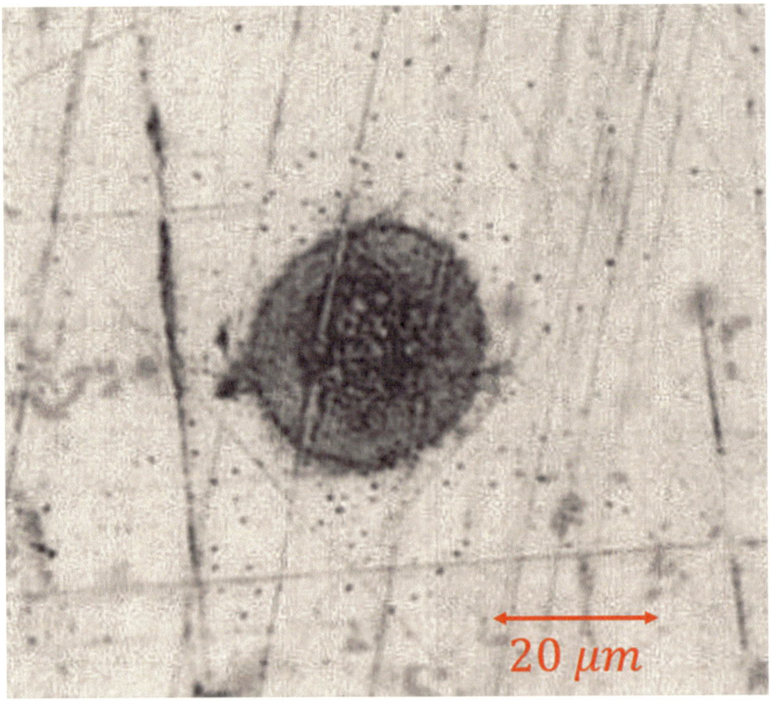

Figure 6.
Example of a spot on bulk Al. In this case, the pulse duration was $\tau = 320 fs$; the pulse energy was $E_P = 18.7$ μJ, and a 20× Mitutoyo Plan Apo NIR B Infinity Corrected Objective was used to obtain the picture (scale is also shown).

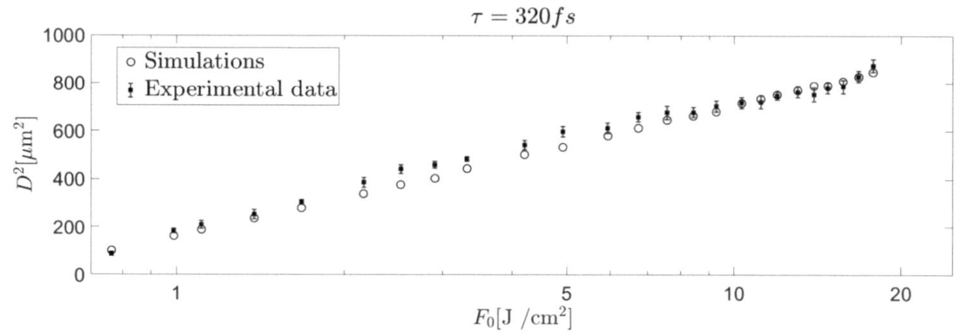

Figure 7.
Comparison between experimentally measured diameters of spots and results from simulations. The experiments were performed using the FemtoLux30 (Ekspla) laser device, which emits in the IR range, $\lambda = 1030$ nm.

3.4.2 Comparison between simulations and experiments

Figure 7 shows the comparison between experimental diameters and results from simulations. The plot shows the square of the diameter against the peak fluence in logarithmic scale, as for this plot a straight line is expected to be obtained for a Gaussian beam according to Liu method [47], which predicts the following fluence dependency of the experimental diameters:

$$D^2 = 2\omega_0^2\, ln\,(F_0) - 2\omega_0^2\, ln\,(F_{\text{thr}}) \tag{40}$$

Where F_{thr} is the threshold fluence for each material.

In general, **Figure 7** shows good agreement between experiments and simulations. The differences between both sets of data remain below 2 μm for all spots, and the average discrepancy between experiments and simulations is under 1 μm.

It should also be mentioned that the discrepancy between experiments and simulations is even smaller for high fluences. This is particularly visible in **Figure 7** above $F_0 = 8$ J/cm^2, where there is an excellent match between both sets of data. This can be explained because phase explosion is especially prevalent at high fluences [35], and in our simulations, we have set the ablation temperature considering the phase explosion condition, see Section 3.2.2. On the other hand, for low fluences, the presence of an initial spallation layer [48] can slightly change the dynamics of ablation, in **Figure 7**, for low pulse energies simulations consistently underestimate the spot radii.

4. Conclusions

In this chapter, we have summarised the main mechanisms which govern ultrashort laser heat transfer and ablation. For long pulses, the heat is transmitted to the target during the laser irradiation, and then usual melting and vaporisation take place; however, for short pulses different physical phenomena appear.

In the case of metals, for pulse durations of a few ps or shorter, there is no time for the lattice of the target to absorb the heat during the pulse. This requires a separate consideration of the evolution of the electron and lattice subsystem via the two-temperature model (TTM). The energy of the pulse is first absorbed by the electron subsystem, which can reach extremely high temperatures of tens of thousands of

Kelvin. Then, the absorbed heat is transferred to the lattice by electron-phonon collisions, until both subsystems reach thermal equilibrium (which usually takes place after a few tens of ps for metals).

The short timescales also produce different ablation mechanisms. For ultrashort pulses, the relevant mechanisms for ablation are phase explosion and, at lower fluences, spallation. Phase explosion is a consequence of the absence of heterogeneous nucleation (the physical phenomena responsible for normal boiling), due to the rapid heating of the lattice. Instead, the molten material is heated well over the boiling point, until it reaches a temperature close to the critical thermodynamic temperature of the material. At this point, homogeneous bubble nucleation takes place, which produces a sudden explosion that is ultimately responsible for ablation.

On the other hand, at low fluences near the ablation threshold, spallation can also take place. This mechanism is a consequence of the isochoric heating of the material, as the timescales will not allow the material to expand. As a result, a high compressive stress will be created, which will evolve into a tensile stress after the interaction with the free surface of the material. Under certain conditions, this tensile stress can cause the ejection of a layer of material. The requirement of isochoric heating limits the pulse durations at which spallation takes place to a few ps at most, while phase explosion can still take place on much longer ns pulses.

In the second part of this chapter, we have presented a novel MATLAB code which simulates fs laser ablation of metals. To perform this task, it solves the TTM equations using finite differences methods. This methodology allows for all the thermal and optical properties of the material to change during the simulation, which is necessary in order to obtain realistic predictions.

We have presented the results obtained for the particular case of aluminium, for which reliable models from previous publications were available for its thermal and optical parameters. The predictions from our code were validated against our own experimental data, obtaining good agreement for the spots' diameters. In future works, this same code can be used to simulate fs laser ablation on other metals or to predict quantities of industrial interest, such as the ablation efficiency.

Acknowledgements

This research has been partially funded by Grant PID2022-1384340B-C54 (SCALING), funded by MICIU/AEI/10.13039/501100011033, by "ERDF/EU" Grant PRE2020-095595 (Sergio Vela), funded by MICIU/AEI/10.13039/501100011033, and by "ESF Investing in your future". The authors also gratefully acknowledge the Universidad Politécnica de Madrid (www.upm.es) for providing computing resources on Magerit Supercomputer.

Conflict of interest

The authors declare no conflict of interest.

Author details

Sergio Vela Liñán[1], Miguel Morales[1,2*], David Munoz-Martin[1,3] and Carlos Molpeceres[1,2]

1 Centro Láser, Universidad Politécnica de Madrid, Madrid, Spain

2 Escuela Técnica Superior de Ingenieros Industriales, Universidad Politécnica de Madrid, Madrid, Spain

3 Escuela Técnica Superior de Ingeniería y Diseño Industrial, Universidad Politécnica de Madrid, Madrid, Spain

*Address all correspondence to: miguel.morales@upm.es

References

[1] Dong X, Liu S, Song H, Peng G. Growth of large-area, few-layer graphene by femtosecond pulsed laser deposition with double-layer ni catalyst. Journal of Materials Science. 2016;**52**(4): 2060-2065. DOI: 10.1007/s10853-016-0494-3

[2] Ali N, Bashir S, Kalsoom U i, Begum N, Rafique MS, Husinsky W. Effect of liquid environment on the titanium surface modification by laser ablation. Applied Surface Science. 2017; **405**:298-307. DOI: 10.1016/j.apsusc.2017.02.047

[3] Gattass RR, Mazur E. Femtosecond laser micromachining in transparent materials. Nature Photonics. 2008;**2**(4): 219-225. DOI: 10.1038/nphoton.2008.47

[4] Chichkov BN, Momma C, Nolte S, Alvensleben F, Tünnermann A. Femtosecond, picosecond and nanosecond laser ablation of solids. Applied Physics A Materials Science and Processing. 1996;**63**(2):109-115. DOI: 10.1007/bf01567637

[5] Redka D, Winter J, Gadelmeier C, Djuranovic A, Glatzel U, Minár J, et al. Control of ultrafast laser ablation efficiency by stress confinement due to strong electron localization in high-entropy alloys. Applied Surface Science. 2022;**594**:153427. DOI: 10.1016/j.apsusc.2022.153427

[6] Poprawe R, Boucke K, Hoffman D. Laser Beams. Berlin Heidelberg: Springer; 2018. pp. 111-139. DOI: 10.1007/978-3-642-01234-1_5

[7] Welty J, Rorrer GL, Foster DG. Fundamentals of Momentum, Heat, and Mass Transfer. Hoboken, New Jersey, U.S.: Wiley; 2014. Available from: https://books.google.es/books?id=PVCfDwAAQBAJ

[8] Kaganov MI, Lifshitz IM, Tanatarov LV. Relaxation between electrons and the crystalline lattice. Soviet Physics JETP. 1957;**4**(2):173-183

[9] Anisimov SI, Kapeliovich BL, Perelman TL. Electron emission from metal surfaces exposed to ultrashort laser pulses. Zhurnal Eksperimentalnoi i Teoreticheskoi Fiziki. 1974;**66**:776-781

[10] Dong Y, Zongpu W, You Y, Yin C, Wenhui Q, Li X. Numerical simulation of multi-pulsed femtosecond laser ablation: Effect of a moving laser focus. Optical Materials Express. 2019;**9**(11): 4194. DOI: 10.1364/ome.9.004194

[11] Momenzadeh L, Evteev AV, Levchenko EV, Belova IV, Murch GE, Sohn YH. Phonon thermal conductivity of f.c.c. cu by molecular dynamics simulation. Defect and Diffusion Forum. 2013;**336**:169-184. DOI: 10.4028/www.scientific.net/ddf.336.169

[12] Demaske B, J, Zhakhovsky VV, Inogamov NA, Oleynik II. Ablation and spallation of gold films irradiated by ultrashort laser pulses. Physical Review B. 2010;**82**(6):064113. DOI: 10.1103/physrevb.82.064113

[13] Bévillon E, Colombier JP, Dutta B, Stoian R. Ab initio nonequilibrium thermodynamic and transport properties of ultrafast laser irradiated 316l stainless steel. The Journal of Physical Chemistry C. 2015;**119**(21):11438-11446. DOI: 10.1021/acs.jpcc.5b02085

[14] Alexopoulou VE, Markopoulos AP. A critical assessment regarding two-temperature models: An investigation of the different forms of two-temperature models, the various ultrashort pulsed laser models and computational methods. Archives of

Computational Methods in Engineering. 2023;**31**(1):93-123. DOI: 10.1007/s11831-023-09974-1

[15] Qiu TQ, Tien CL. Heat transfer mechanisms during short-pulse laser heating of metals. Journal of Heat Transfer. 1993;**115**(4):835-841. DOI: 10.1115/1.2911377

[16] Ashcroft NW, Mermin N. Solid State Physics. Florence, KY: Brooks/Cole; 1976

[17] Etchegoin PG, Le Ru EC, Meyer M. An analytic model for the optical properties of gold. The Journal of Chemical Physics. 2006;**125**(16):164705. DOI: 10.1063/1.2360270

[18] Ren Y, Chen JK, Zhang Y. Optical properties and thermal response of copper films induced by ultrashort-pulsed lasers. Journal of Applied Physics. 2011;**110**(11):113102. DOI: 10.1063/1.3662897

[19] Fisher D, Fraenkel M, Henis Z, Moshe E, Eliezer S. Interband and intraband (drude) contributions to femtosecond laser absorption in aluminum. Physical Review E. 2001;**65**(1):016409. DOI: 10.1103/physreve.65.016409

[20] Shikne R, Yoneda H, Chaurasia S. Estimation of electron temperature dynamics in warm dense gold using measured and dft-calculated plasma frequency. Physica Scripta. 2023;**98**(4):045813. DOI: 10.1088/1402-4896/acc0af

[21] Hohlfeld J, Müller JG, Wellershoff S-S, Matthias E. Time-resolved thermoreflectivity of thin gold films and its dependence on film thickness. Applied Physics B: Lasers and Optics. 1997;**64**(3):387-390. DOI: 10.1007/s003400050189

[22] Byskov-Nielsen J, Savolainen J-M, Christensen MS, Balling P. Ultra-short pulse laser ablation of copper, silver and tungsten: Experimental data and two-temperature model simulations. Applied Physics A. 2011;**103**(2):447-453. DOI: 10.1007/s00339-011-6363-7

[23] Verhoeven JCJ, Jansen JKM, Mattheij RMM, Smith WR. Modelling laser induced melting. Mathematical and Computer Modelling. 2003;**37**(3–4):419-437. DOI: 10.1016/s0895-7177(03)00017-7

[24] Shen ZH, Zhang SY, Lu J, Ni XW. Mathematical modeling of laser induced heating and melting in solids. Optics and Laser Technology. 2001;**33**(8):533-537. DOI: 10.1016/s0030-3992(01)00005-6

[25] Allmen MV, Blatter A. Laser Materials Interactions. Boston, MA: Springer US; 2008. pp. 34-65. DOI: 10.1007/978-0-387-72344-0_2

[26] Shugaev MV, He M, Levy Y, Mazzi A, Miotello A, Bulgakova NM, et al. Laser-Induced Thermal Processes: Heat Transfer, Generation of Stresses, Melting and Solidification, Vaporization, and Phase Explosion. Berlin/Heidelberg, Germany: Springer International Publishing; 2021. pp. 83-163. DOI: 10.1007/978-3-030-63647-0_11

[27] Mulser P, Sigel R, Witkowski S. Plasma production by laser. Physics Reports. 1973;**6**(3):187-239. DOI: 10.1016/0370-1573(73)90005-7

[28] Eidmann K. Plasmas at Solid State Density Generated by Ultra-Short Laser Pulses. New York, NY, USA: Springer US; 2001. pp. 315-326. DOI: 10.1007/978-1-4615-1351-3_17

[29] Higginson A, Gray RJ, King M, et al. Near-100 mev protons via a laser-driven transparency-enhanced hybrid acceleration scheme. Nature

Communications. 2018;**9**:724. DOI: 10.1038/s41467-018-03063-9

[30] Remington BA, David Arnett R, Paul D, Takabe H. Modeling astrophysical phenomena in the laboratory with intense lasers. Science. 1999;**284**(5419):1488-1493. DOI: 10.1126/science.284.5419.1488

[31] Leveugle E, Ivanov DS, Zhigilei LV. Photomechanical spallation of molecular and metal targets: Molecular dynamics study. Applied Physics A. 2004;**79**(7): 1643-1655. DOI: 10.1007/s00339-004-2682-2

[32] Miotello A, Kelly R. Critical assessment of thermal models for laser sputtering at high fluences. Applied Physics Letters. 1995;**67**(24):3535-3537. DOI: 10.1063/1.114912

[33] Kelly R, Miotello A. Comments on explosive mechanisms of laser sputtering. Applied Surface Science. 1996;**96–98**:205-215. DOI: 10.1016/0169-4332(95)00481-5

[34] Ivanov DS, Zhigilei LV. Combined atomistic-continuum modeling of short-pulse laser melting and disintegration of metal films. Physical Review B. 2003;**68**(6):064114. DOI: 10.1103/physrevb.68.064114

[35] Zhigilei LV, Lin Z, Ivanov DS. Atomistic modeling of short pulse laser ablation of metals: Connections between melting, spallation, and phase explosion. The Journal of Physical Chemistry C. 2009;**113**(27):11892-11906. DOI: 10.1021/jp902294m

[36] Vela S, Morales M, Munoz-Martin D, Molpeceres C. Numerical simulation of femtosecond laser ablation of aluminium. Results in Physics. 2025;**75**: 108313. DOI: 10.1016/j.rinp.2025.108313

[37] Fornberg B. Generation of finite difference formulas on arbitrarily spaced grids. Mathematics of Computation. 1988;**51**(184):699-706

[38] Zhou T, Hong Y, Fang Z, Perrie W, Fei Y, Youyou H, et al. Experimental and numerical study of multi-pulse picosecond laser ablation on 316l stainless steel. Optics Express. 2023; **31**(23):38715. DOI: 10.1364/oe.505324

[39] Morel V, Bultel A, Chéron BG. The critical temperature of aluminum. International Journal of Thermophysics. 2009;**30**(6):1853-1863. DOI: 10.1007/s10765-009-0671-6

[40] Vela S, Molpeceres C, Morales M. Development of a new analytical solution for electronic heat capacity for higher electron temperatures. Results in Physics. 2024;**59**:107565. DOI: 10.1016/j.rinp.2024.107565

[41] Winter J, Redka D, Minar J, Schmidt M, Huber H. Resolving transient temperature and density during ultrafast laser ablation of aluminum. Applied Physics A. 2023;**129**:09. DOI: 10.1007/s00339-023-06922-5

[42] Ziman JM. Principles of the Theory of Solids. 2nd ed. Cambridge, England: Cambridge University Press; 2013

[43] Petrov YV, Inogamov NA, Khokhlov VA, Migdal KP. Electron thermal conductivity of nickel and aluminum in solid and liquid phases in two-temperature states. Journal of Physics Conference Series. 2021;**1787**(1): 012025

[44] Majchrzak E, Poteralska J. Two-temperature microscale heat transfer model, part II: Determination of lattice parameters. Scientific Research of the Institute of Mathematics and Computer Science. 2010;**9**:109-119

[45] Pallik E. Handbook of Optical Constants of Solids. San Diego: Academic Press; 1985

[46] Vial A, Laroche T. Comparison of gold and silver dispersion laws suitable for FDTD simulations. Applied Physics B: Lasers and Optics. 2008;**93**(1):139-143

[47] Liu JM. Simple technique for measurements of pulsed gaussian-beam spot sizes. Optics Letters. 1982;7(5):196. DOI: 10.1364/ol.7.000196

[48] Winter J, Rapp S, Spellauge M, Eulenkamp C, Schmidt M, Huber HP. Ultrafast pump-probe ellipsometry and microscopy reveal the surface dynamics of femtosecond laser ablation of aluminium and stainless steel. Applied Surface Science. 2020;**511**: 145514. DOI: 10.1016/j.apsusc.2020.145514

Chapter 11

Using Q-Switched Lasers for Multi-Beam Multi-Target Pulsed Laser Deposition of Photonic High-Entropy Alloy Films

Abdalla M. Darwish, Sergey S. Sarkisov and Darayas N. Patel

Abstract

The chapter describes the application of Q-switched lasers to the concurrent multi-beam multi-target pulsed laser deposition (CMBMT-PLD) of the high-entropy alloy (HEA) films. A six-beam system for such a process additionally included permanent magnets under the PLD targets and the substrate and a substrate heater. The magnets narrowed down the plasma plumes from the targets and increased the material deposition rate. Three- (Au-Ag-Cu), four- (previous three + Al), five- (previous four + Ti), and six-metal (previous five + W) films were deposited on Cu, Al, Fe, and Ni substrates. Scanning electron microscopy and X-ray analysis indicated that the films were the aggregates of crystalline nanoparticles (NPs) of mostly individual metals and double-metal alloys. Heating the substrates to up to 300°C resulted in the formation of a wider variety of double-alloy nanoparticles due to higher mobility and reactivity of the deposited atomic species. The HEA films amplified the photoluminescence (PL) of the perovskite $CsPbBr_3$ quantum dots (QDs) deposited on the top by 6 to 16 times. This was because of two simultaneous effects. The first was the enhancement of the incident ultraviolet radiation (372-nm wavelength laser) by the localized surface plasmon polariton resonance in the metal NPs surrounding the perovskite QD. The second was the fast decay of the excited state of the QD due to a resonant coupling between the QD and the NPs in its vicinity. The findings described in this chapter might be useful for the applications in optical sensing.

Keywords: pulsed laser deposition with magnetic field, metal plasmonic nanostructures, metal alloy plasmonic films, photoluminescence amplification, surface plasmon polariton resonance

1. Introduction

Pulsed laser deposition (PLD) has emerged as a versatile physical vapor deposition technique for synthesizing high-quality thin films with precise stoichiometric control. In this process, a high-energy laser ablates a target material, generating a plasma plume that deposits onto a substrate under controlled vacuum or gas environments. The technique's effectiveness depends critically on laser parameters, including pulse duration, energy density, and repetition rate, which govern ablation dynamics and film growth mechanisms.

The integration of Q-switched lasers in PLD represents a significant advancement in the fabrication of thin films, leveraging the unique properties of Q-switched lasers to enhance deposition efficiency and film quality. Q-switched lasers, known for their ability to produce high-energy (in the order of 1 J/pulse), short-duration (1 to 10 ns) pulses, facilitate the ablation of target materials, resulting in the deposition of thin films with the tailored properties [1]. The temporal characteristics of Q-switched outputs facilitate rapid vaporization of target surfaces through electronic excitation rather than thermal mechanisms, preserving material stoichiometry during transfer to substrates [2].

Further development of Q-switched laser-assisted PLD is the concurrent multi-beam multi-target pulsed laser deposition (CMBMT-PLD) [3–5]. The principle of such a method is illustrated in **Figure 1**. Multiple targets (three targets in **Figure 1**) are concurrently ablated by three beams from one or several synchronized Q-switched lasers. The material plasma plumes ejected from the targets overlap on a substrate, condense, and make up a nanocomposite film. Such films can be made of materials with quite different physical and chemical properties at various controlled proportions.

As a further advancement, PLD in a magnetic field has emerged for enhancing the properties of thin films and nanostructures. This method not only improves the deposition rate but also influences the morphology and orientation of the resulting materials. A novel approach using PLD in a magnetic field has been

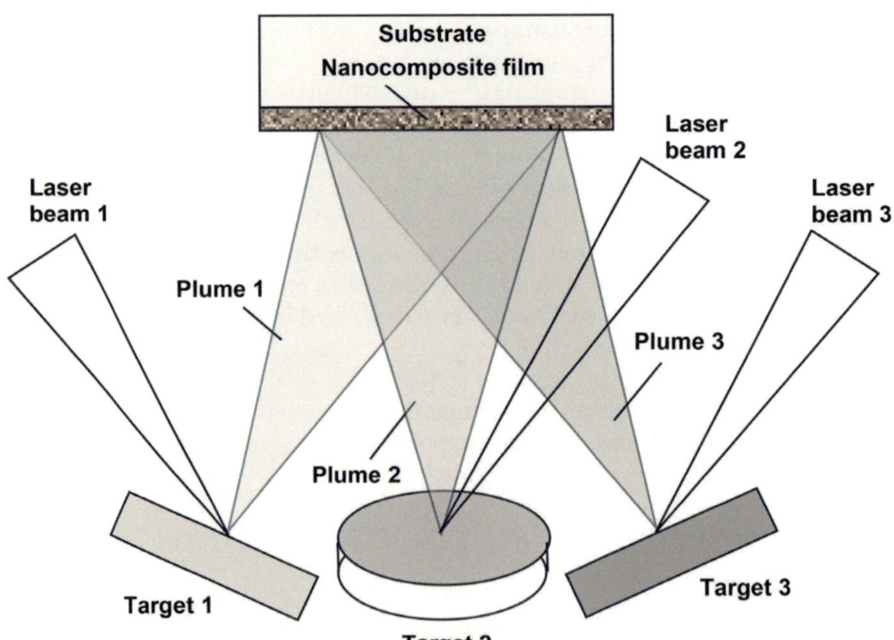

Figure 1.
The schematic diagram illustrating the concurrent multi-beam multi-target pulsed laser deposition (CMBMT-PLD) setup.

demonstrated to fabricate oriented composite nanowires from iron and silver, achieving alignment along magnetic force lines [6]. The deposition of diamond-like carbon films under an inhomogeneous magnetic field resulted in a 1.7-fold increase in deposition rate and improved surface morphology, with reduced roughness and enhanced mechanical properties [7]. A specific PLD method for doped yttrium iron garnet (YIG) thin films has been established, achieving high magneto-optical performance, crucial for integrated optical devices [8].

In this chapter, we describe our efforts toward using CMBMT-PLD without and with a magnetic field for making high-entropy alloy (HEA) plasmonic films. High-entropy alloys (HEAs) are a groundbreaking class of materials that redefine traditional alloy design by combining three to six or more principal elements in near-equal atomic proportions. This multi-principal-element approach creates a high configurational entropy, stabilizing single-phase solid solutions rather than intermetallic compounds, despite the complex atomic interactions [9]. Plasmonic high-entropy materials represent a subclass of HEAs that combines tunable optical properties with extreme durability. Plasmonic HEAs are composed of noble metals Au, Ag, and Cu; transition metals (typically from groups 4–6: Ti, Zr, Hf, V, Nb, Ta, Cr, Mo, W); and other metals (Al, Ni, Fe, Sn) [10]. While the term "high-entropy alloys" typically refers to metallic systems, they also include plasmonic high-entropy carbides (PHECs)—ceramic materials composed of five or more transition metals bonded with carbon [11]. Plasmonic HEA films demonstrate high refractive indices and strong plasmonic resonances, which are beneficial for applications like high-contrast gratings [12]. However, there were no reports on the enhancement of photoluminescence (PL) by metallic HEA films. Magnetron sputtering is commonly used to deposit HEA films, influencing their microstructure and enhancing properties like oxidation resistance [13, 14]. Techniques like 3D printing and femtosecond laser ablation are employed to create HEA structures, allowing for precise control over dimensions and properties [10, 15].

We report, for the first time to our knowledge, on the enhancement of PL by the metallic HEA films made by the CMBMT-PLD method. In Section 2, we describe the materials used; in Section 3, the experimental setup for filmmaking; and in Section 4, results and discussion, followed by conclusions.

2. Materials

We used as PLD targets the 2.54-cm-diameter metal targets for magnetron sputtering from Kurt J. Lesker Company [16]. The selected metals were Au, Ag, Cu, Al, Ti, and W. The substrates for the HEA films were made of Cu, Al, Fe, and Ni.

The property of the plasmonic HEA films to enhance PL was investigated using homemade perovskite $CsPbBr_3$ quantum dots (QDs), demonstrating PL in the visible region under excitation with a UV source. For making perovskite QDs, we used the following synthetic route [17]. All the chemicals were obtained from Millipore-Sigma. First, we dissolved 0.85 g of CsBr in 100 ml of dimethylformamide (DMF) and stirred at 140°C to form cesium precursor solution. Then, we dissolved 0.37 g of $PbBr_2$ in 100 ml of DMF and stirred at 140°C to form lead precursor solution. After dissolving $PbBr_2$, we added to the lead precursor solution 1 ml of oleic acid (OA) and 1 ml of oleylamine (OLA). During the synthesis, we quickly injected the cesium precursor solution into the lead precursor mixture to initiate nucleation of Cs-Pb-bromide nanocrystals. After 5 min of stirring at 140°C, the mixture was placed in an ice bath to cool down and stop further formation of nanocrystals. By adding the antisolvent hexane, Cs-Pb-bromide nanocrystals were caused to precipitate. The resulting mixture was centrifuged at 3500 rpm and washed with hexane several times to remove unreacted products. The purified QDs (nanocrystals) were dried under vacuum and stored in an inert atmosphere of nitrogen. The size of the obtained QDs was determined using the dynamic light scattering (DLS) method as (148 ± 25) nm. They emitted green light with a peak at ~514 nm being excited with ultraviolet (UV) radiation.

3. Experimental setup

For making HEA films, we built a CMBMT-PLD system with a vacuum chamber that has six ports for laser beams and six target holders. The front view of the system is presented in **Figure 2**. The vacuum chamber is mounted on a movable table with the control instrumentation beneath. On the top of the chamber, there are sets of mirrors

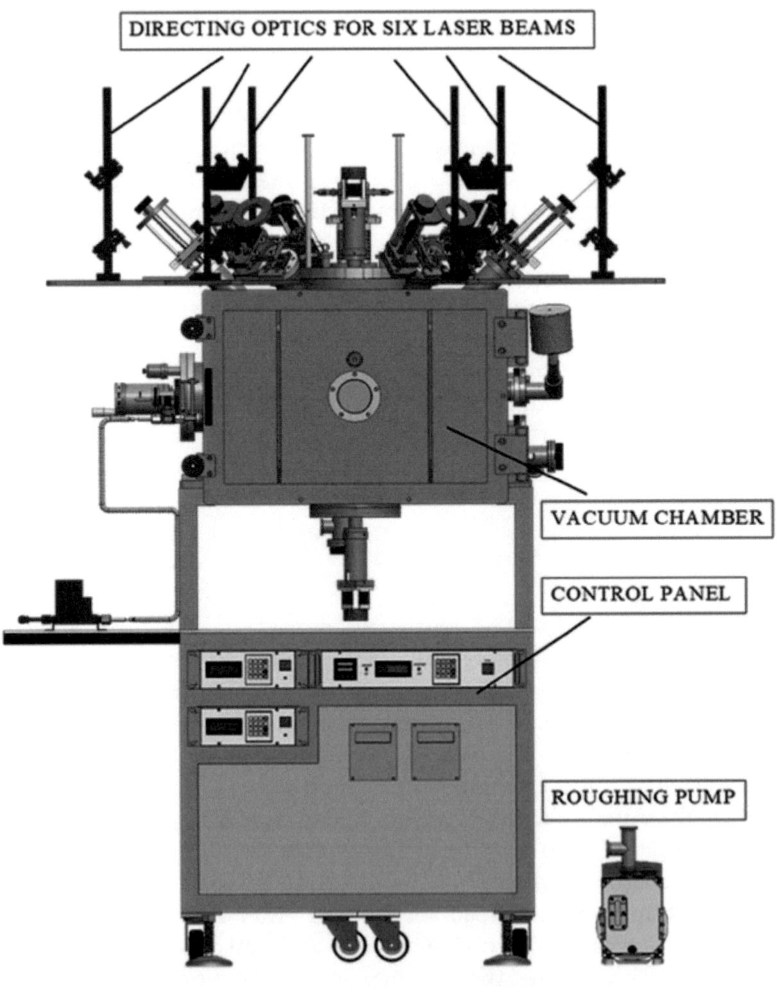

Figure 2.
Front view of the CMBMT-PLD system.

DOI: http://dx.doi.org/10.5772/intechopen.1011076

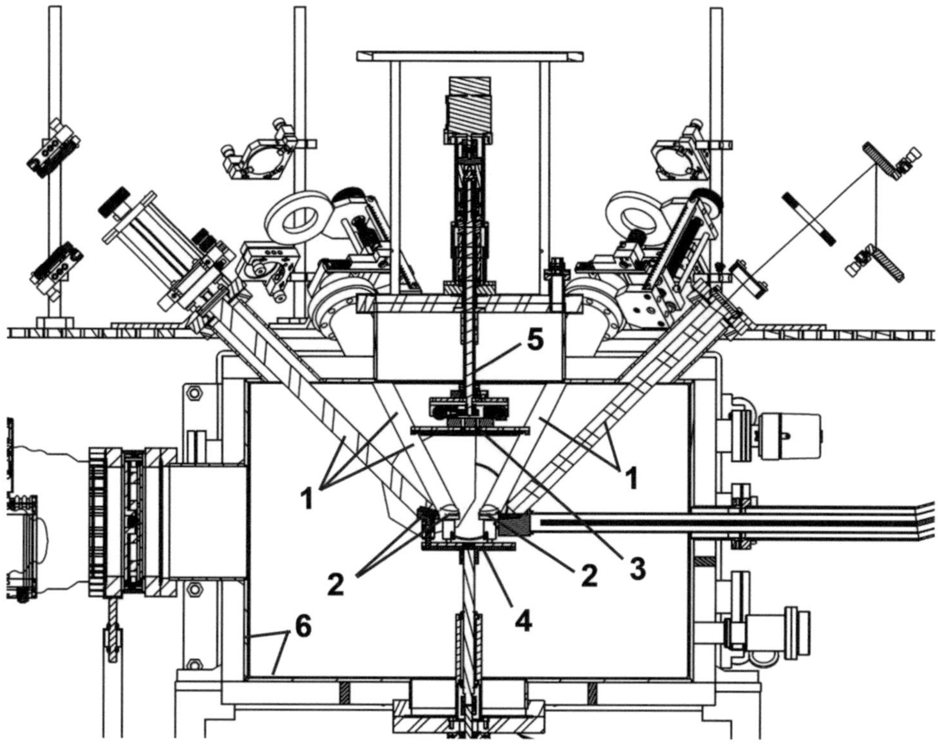

Figure 3.
The schematic view of the interior of the PLD vacuum chamber. 1: laser beams; 2: targets; 3: substrate; 4: target holder; 5: substrate holder; 6: chamber walls.

to direct six laser beams inside through the attached windows. The interior of the PLD chamber is schematically presented in **Figure 3**. Six targets are mounted on a target holder at the bottom of the chamber. Six laser beams are directed to the targets at ~60° incidence angle. The substrate holder is attached to the top. It can rotate and has a heater. **Figure 4** shows the location of permanent magnets under the targets and the substrate and the direction of their magnetic fields in the case of PLD with a magnetic field. The magnetic fields of the magnets under the targets are directed toward the center of the substrate at an angle of 67° with respect to the vertical direction. The magnetic field of the magnet under the substrate is vertical.

4. PLD process

The PLD process was conducted with three to six beams from synchronized Q-switched lasers with a pulse duration of ~5 ns. Three- (Au + Ag + Cu), four- (Au + Ag + Cu + Al), five- (Au + Ag + Cu + Al + Ti), and six-metal (Au + Ag + Cu + Al + Ti + W) HEA films were deposited on Cu, Al, Fe, and Ni substrates cut in the form of 2.54 × 2.54 cm (1 × 1 inch) squares.

The Au target was ablated with a frequency-doubled Q-switched Nd:YAG laser (532 nm wavelength) at a pulse repetition rate of 10 pulses per second. The average power of the laser beam outside the chamber was 1.9 W, and inside, it was 1.5 W.

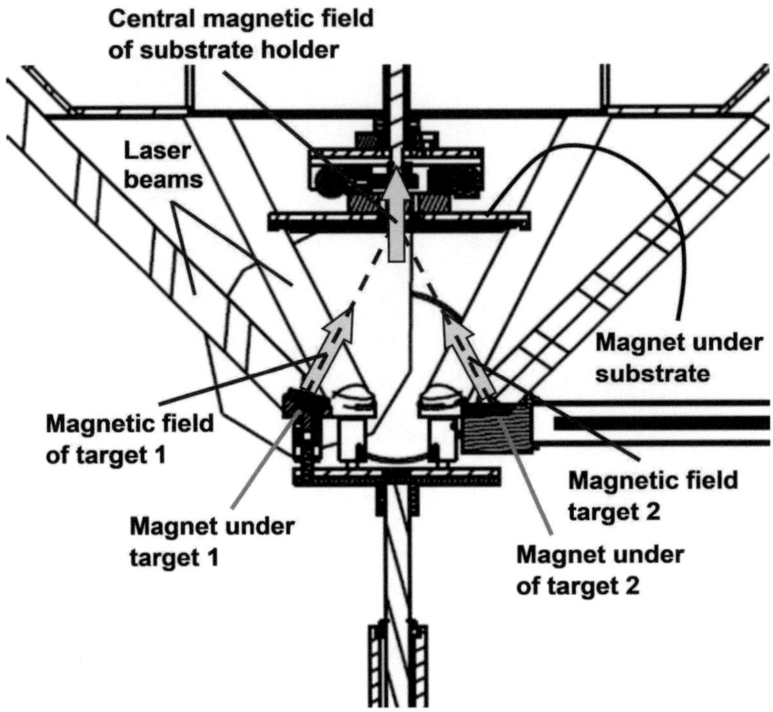

Figure 4.
Schematic view of the disposition of magnets in the PLD vacuum chamber. Gray arrows show the direction of the magnetic field of the permanent magnets mounted under the PLD targets and the substrate.

Laser beam diameter on the target was ~4.4 mm (0.151 cm^2 beam area), and pulse energy was 0.15 J per pulse (0.99 J/cm^2 energy density). The Ag target was ablated with a frequency-doubled Q-switched Nd:YAG laser (532 nm wavelength) at a pulse repetition rate of 50 pulses per second. The average power of the laser beam outside the chamber was 8.5 W, and inside, it was 5.5 W. Laser beam diameter on the target was ~5.5 mm (0.234-cm^2 beam area), and pulse energy was 0.11 J per pulse (0.47 J/cm^2 energy density). The Cu target was ablated with a frequency-doubled Q-switched Nd:YAG laser (532-nm wavelength) at a pulse repetition rate of 10 pulses per second. The average power of the laser beam outside the chamber was 2.5 W, and inside, it was 1.7 W. Laser beam diameter on the target was ~3.7 mm (0.107 cm^2 beam area), and pulse energy was 0.17 J per pulse (1.59 J/cm^2 energy density). The Al target was ablated with a frequency-doubled Q-switched Nd:YAG laser (532-nm wavelength) at a pulse repetition rate of 50 pulses per second. The average power of the laser beam outside the chamber was 7.1 W, and inside, it was 5.2 W. Laser beam diameter on the target was ~4.1 mm (0.13 cm^2 beam area), and pulse energy was 0.104 J per pulse (0.8 J/cm^2 energy density). The Ti target was ablated with a frequency-doubled Q-switched Nd:YAG laser (532-nm wavelength) at a pulse repetition rate of 50 pulses per second. The average power of the laser beam outside the chamber was 2.9 W, and inside, it was 1.9 W. Laser beam diameter on the target was ~3.8 mm (0.112-cm^2 beam area), and pulse energy was 0.038 J per pulse (0.34 J/cm^2 energy density). The W target was ablated with a frequency-doubled Q-switched Nd:YAG laser (532-nm wavelength) at a pulse repetition rate of 10 pulses per second. The average power of the laser beam outside the chamber was 1.3 W, and inside, it was 1.2 W. Laser beam diameter on the target was ~3.2 mm (0.08-cm^2 beam area), and pulse energy was 0.102 J per pulse (1.27 J/cm^2 energy density). All targets were tilted at an angle of 25.60° with respect to a horizontal plane. The vacuum inside the chamber was kept at 3.8 × 10^{-6} Torr. Rastering of the laser beams over the targets was with an angle range of 5 degrees, with a speed of 2 degrees per second. The substrates were rotated at an angular speed of 2 rpm. The substrates during deposition were kept at room temperature or heated to 300°C. In the case of PLD with a magnetic field, the average magnetic field in the center of the chamber was ~334 mT. The duration of the PLD process was 2 hours.

Magnetic field affected the shapes of the plasma plumes ejected from the targets by the laser beams during the CMBMT-PLD process, as shown in **Figure 5**. The plasma plumes for the process with a magnetic field were narrower than without the field. In PLD, plasma plumes are made of positive ions of the ejected material and are expanding due to electrostatic repulsion between the ions. Narrowing of the plume

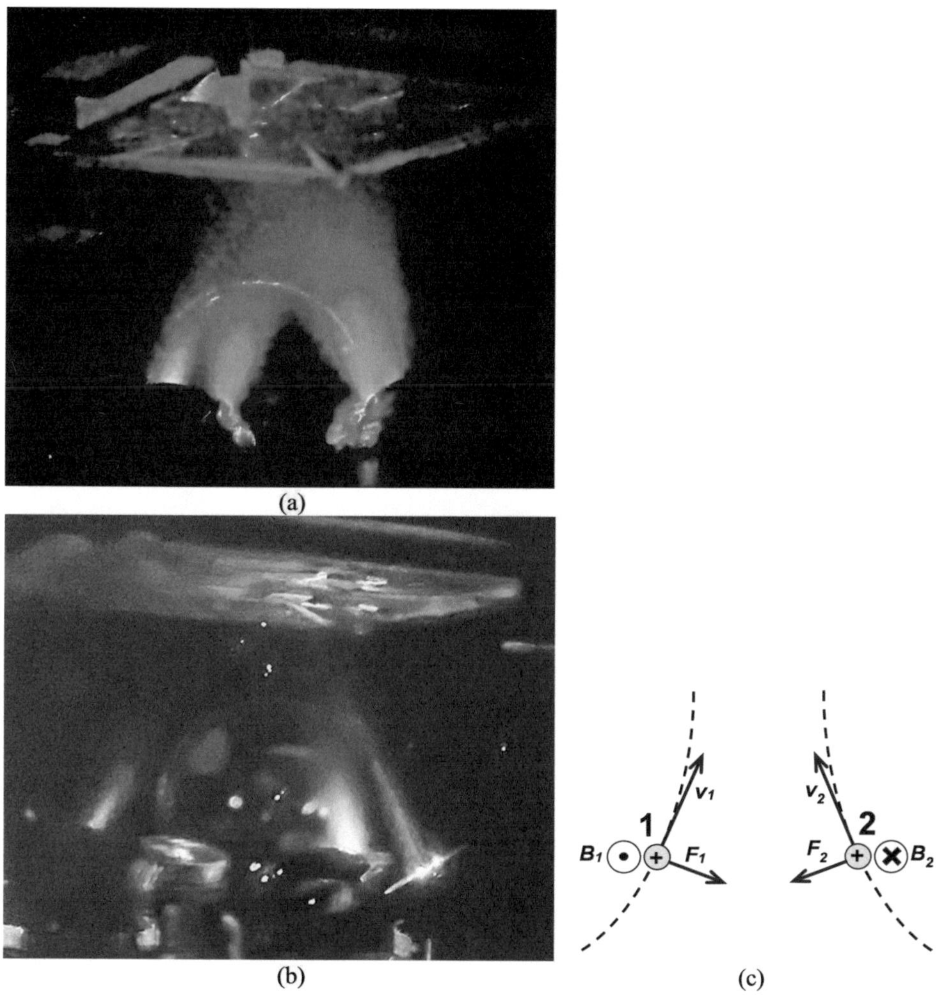

Figure 5.
Photographs of the plasma plumes propagating from the laser-ablated targets toward the substrate during the CMBMT-PLD process without (a) and with the magnetic field (b). Sketch explaining the narrowing of the plasma plume in the case of a magnetic field (c).

in the case of a magnetic field can be explained using the schematic in **Figure 5c**. Positive ions 1 and 2 with electric charge q are moving with velocities $\mathbf{v}_1$ and $\mathbf{v}_2$ on the left and right edges of the plasma plume, respectively. Horizontal component $\mathbf{B}_1$ of the magnetic field near ion 1 is created by the magnet sitting under the adjacent target to the left. It is directed upwards, perpendicular to the plane of the sketch. Lorentz force $\mathbf{F}_1 = q[\mathbf{v}_1 \times \mathbf{B}_1]$ (brackets denote vector product) pushes ion 1 to the right. Correspondingly, the horizontal component $\mathbf{B}_2$ of the magnetic field created by the magnet under the adjacent target to the right is directed downwards, perpendicular to the plane of the sketch. The Lorentz force $\mathbf{F}_2 = q[\mathbf{v}_2 \times \mathbf{B}_2]$ pushes ion 2 to the left. The overall result would be bringing the ions in the left and right parts of the plume close to each other and narrowing the plume.

5. HEA films

The examples of the HEA films produced are presented in **Figure 6**. The six-metal HEA films absorb more visible light and look darker. The properties of the films were analyzed using scanning electron microscopy (SEM), accompanied by energy-dispersive X-ray spectroscopy (EDS), X-ray diffraction (XRD) spectroscopy, and optical diffuse reflectance spectroscopy. **Figure 7** presents a typical SEM image of a four-metal (Au, Ag, Cu, and Al) film on a cold (at room temperature) Fe substrate. Regardless of the number of metals, type of substrate, its temperature (room temperature or 300°C), and the lack or presence of a magnetic field, the SEM images looked similar. The films were the aggregates of nanoparticles (NPs) with sizes from 50 to 10,000 nm. **Figure 8** shows the EDS spectrum of the HEA film in **Figure 7**. EDS was taken with a JED-2300

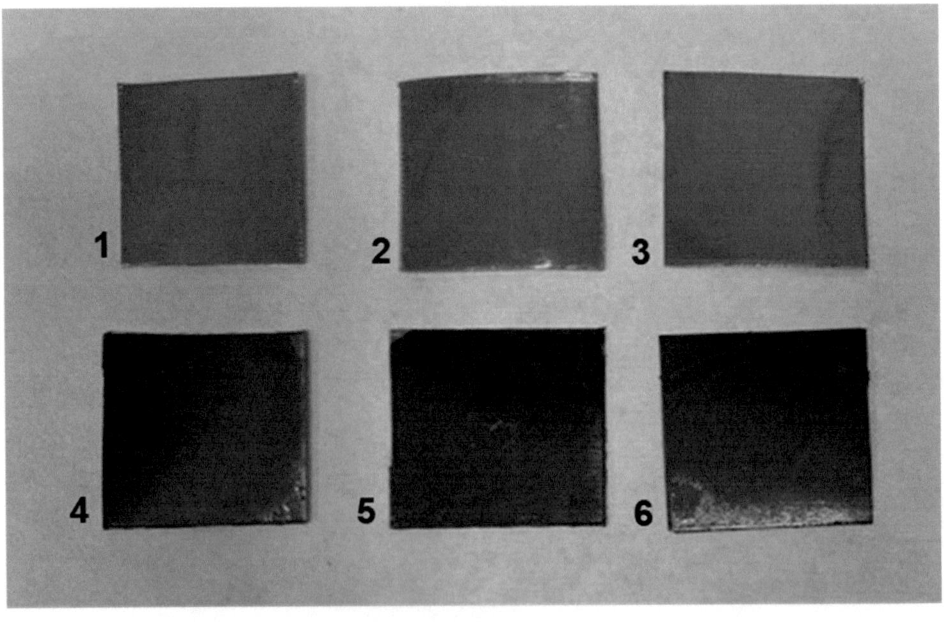

Figure 6.
Photographs of the samples of HEA films deposited on cold substrates without a magnetic field. 1: Au + Ag + Cu on Cu substrate; 2: Au + Ag + Cu on Al; 3: Au + Ag + Cu on Fe; 4: Au + Ag + Cu + Al + Ti + W on Cu; 5: Au + Ag + Cu + Al + Ti + W on Al; and 6: Au + Ag + Cu + Al + Ti + W on Fe.

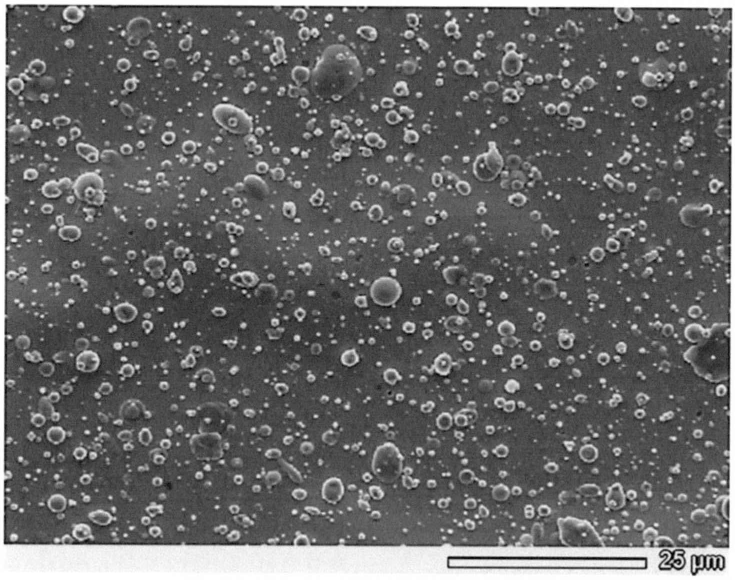

Figure 7.
Example of a scanning electron microscope (SEM) image of a four-metal (Au, Ag, Cu, and Al) film on an Fe substrate.

Analysis Station from Japan Electron Optics Laboratory (JEOL). The spectrum is typical for the EDS of the deposited films, regardless of the number of metals and the deposition conditions. The X-ray spectral peaks of Au, Ag, and Cu were reliably identified in all the samples. But EDS faced challenges in identifying the spectra of aluminum, titanium, and tungsten due to the following several factors. (A) Overlapping peaks: The X-ray emission lines of these elements could overlap with those of other elements,

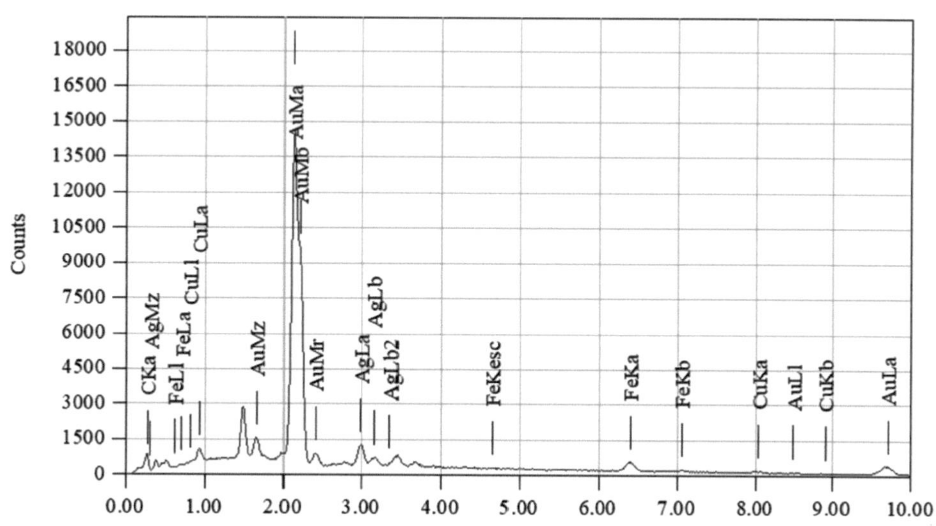

Figure 8.
EDS spectrum of the film in **Figure 7**. *X-ray peaks were attributed to Au, Ag, Cu, and Fe.*

making it difficult to distinguish them accurately. For example, aluminum's Kα peak could overlap with silicon's Kα peak, which is common in many materials. (B) Low-energy X-rays: Aluminum and titanium emit low-energy X-rays, which are more likely to be absorbed by the detector window or the sample itself, reducing the signal strength and making detection harder. (C) High atomic number effects: Tungsten, being a high atomic number element (74 atomic number, 183 atomic mass), produced intense X-ray peaks that could overshadow the peaks of lighter elements in the sample, complicating the analysis.

EDS, with its deficiencies, was backed up by XRD analysis of the films with a D2 Phaser spectrometer from Bruker. Typical XRD spectra (their fragments) of the HEA films (five-metal films on aluminum substrate in this case) deposited without and with a magnetic field are presented in **Figures 9** and **10**, respectively. In the XRD spectrum of the HEA film deposited on the cold substrate without a magnetic field, seven spectral crystalline patterns could be identified: cubic metals Au, Ag, Al, and Ti, and cubic compounds Au_3Cu (Bogdanivite), $Ag_{0.985}Cu_{0.005}$, and AuAg. The spectrum of the film deposited on the hot substrate with a magnetic field has a richer collection (ten) of crystalline patterns: cubic metals Au, Ag, Al, and Ti; cubic compounds Au_3Cu (Bogdanovite), $Ag_{0.842}Al_{0.158}$, and $Ag_{3.88}Cu_{0.12}$; tetragonal compounds $CuTi_3$ and AlTi; and orthorhombic compounds $Al_{2.5}Ti_{1.5}$. These spectral patterns are marked in **Figures 9** and **10** by vertical lines of different shades of gray. There are more of these lines in the spectrum in **Figure 10**. It is important to mention that the film deposited on the hot substrate had a wider variety of identified alloys (mostly double-metal ones). A similar tendency was observed for other films. This can be explained by a higher mobility and reactivity of the atom species condensed on the hot substrate and a greater concentration of the species in the plumes narrowed down

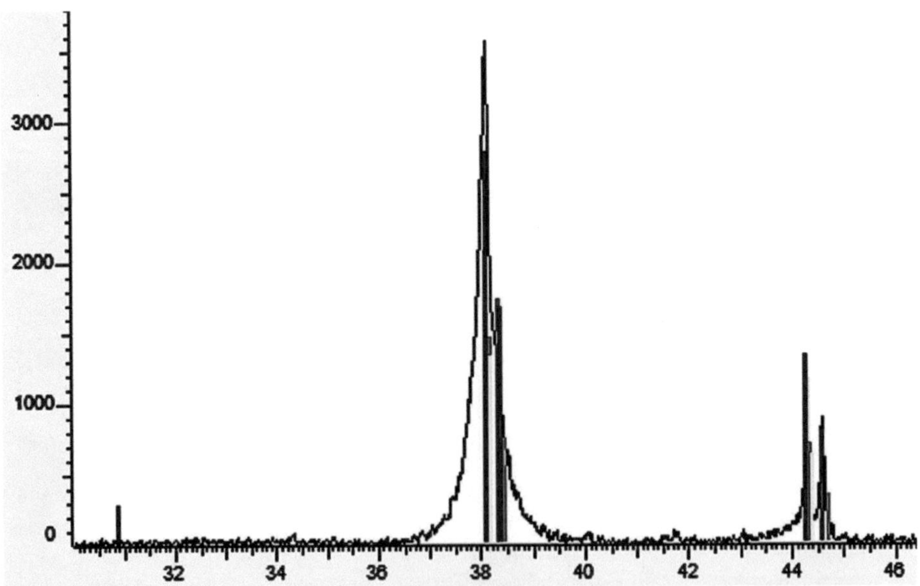

Figure 9.
Fragment of the XRD spectrum of five-metal (Au + Ag + Cu + Al + Ti) HEA film deposited on cold (at room temperature) Al substrate without a magnetic field. The quantity of the horizontal axis is the diffraction angle 2 Theta in degrees. The quantity of the vertical axis is the intensity of the diffracted X-rays in counts. Vertical bars of different shades of gray represented the computed XRD spectral lines of various metals and alloys described in the text.

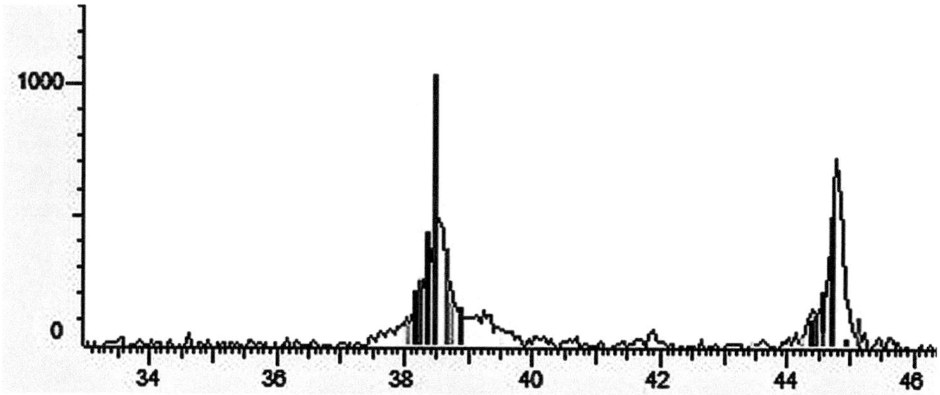

Figure 10.
Fragment of the XRD spectrum of five-metal (Au + Ag + Cu + Al + Ti) HEA film deposited on heated (at 300°C) Al substrate with magnetic field. The quantity of the horizontal axis is the diffraction angle 2 Theta in degrees. The quantity of the vertical axis is the intensity in counts.

by the magnetic field upon their arrival at the substrate. A specific feature of the films deposited on both cold and hot substrates is the presence of the diffraction peaks of the Bogdanovite alloy Au_3Cu in their XRD spectra. Based on the XRD data, it can be assumed that heating the substrate to higher temperatures will result in the formation of more complex compounds (triple- and more-metal alloys) in the films. At the present state, the obtained films were mostly the aggregates of either single-metal (Ag, Au, Cu, etc.) or double-metal alloy NPs.

Optical diffuse reflectance spectroscopy of the films was conducted with a Shimadzu UV-VIS spectrophotometer UV-2600i Plus with an integrating sphere.

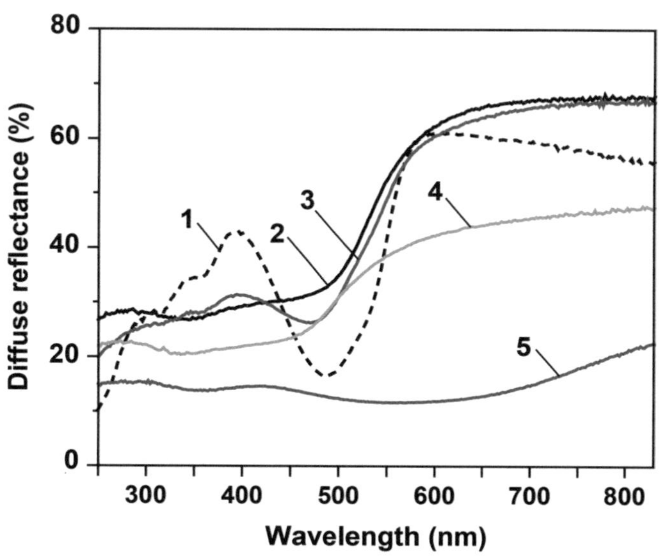

Figure 11.
Diffuse reflectance spectra of the Cu substrate (curve 1), triple metal (Au + Ag + Cu) HEA film on Cu substrate (2), four-metal (Au + Ag + Cu + Al) film on Cu (3), five-metal (Au + Ag + Cu + Al + Ti) film on Cu (4), and six-metal (Au + Ag + Cu + Al + Ti + W) film on Cu (5).

The spectroscopy was useful in getting the information on masking a substrate with the deposited film, especially in the case of a copper substrate. Typical diffuse reflectance spectra are presented in **Figure 11**. The spectrum of the copper substrate (dashed curve 1) has a characteristic dip at around 488 nm, explaining the yellow-red color of copper. This dip is visible in the tri- and four-metal films (curves 2 and 3) but becomes indistinguishable in the six-metal film. The latter film had such an amount of deposited material that it masked the substrate from incident light.

6. Amplification of photoluminescence

We investigated the application of the fabricated HEA films to amplify photoluminescence (PL) of the exemplary samples of perovskite $CsPbBr_3$ QDs excited with an ultraviolet (UV) source. Perovskites find their ever-increasing use in sensors and photovoltaic cells [18]. Integration of various light emitters with plasmonic nanostructures has emerged as an innovative pathway for applications in quantum technologies, sensing, and imaging [18–28]. The emission of light can be enhanced in the presence of metallic nanostructures that sustain localized surface plasmon polaritons (LSPPs). The plasmonic modes intensify and localize electromagnetic fields within nanoscale volumes, overcoming the diffraction limit of light and enhancing the excitation of the emitting species (atoms, molecules, nanoparticles, and QDs) placed in the proximity of the metallic surface. Additionally, plasmonic modes in metallic nanostructures also accelerate the spontaneous emission rate of the emitters (Purcell effect) due to the introduction of new radiative and non-radiative decay channels, in other words, an enhanced local density of photonic states. Further, these nanostructures can modify the radiation pattern of the emitter, enabling a more directional emission as compared to that of the isolated one.

The samples for the PL amplification experiments were prepared using the following procedure. Perovskite QDs described in Section 2 were dispersed in toluene at a concentration of 0.03 g solids to 10 ml liquid. The suspension was spin-cast on microscope cover glass slides (~0.13 mm thickness), and the metal substrates were coated with the HEA films. Spin casting was conducted at a speed of 1000 rpm for 60 s. All the perovskite-coated samples were dried for 2 h on a hot plate at 65°C. The perovskite QDs were excited with a 372-nm UV diode laser LBX-375-70-CSB-PPA from Oxxius (laser beam diameter was 5 mm and average power was 3.5 mW). The PL spectra of the excited QDs were measured using an Avantes optical fiber-coupled spectrometer, AvaSpec-ULS4096CL-EVO, with an integrating sphere, AvaSphere-30. Two excitation schemes were used. In the first scheme, the laser beam excited the perovskite QDs deposited on glass slides. The glass slides were mounted on top of bare HEA films. In the second scheme, the laser beam excited the QDs deposited directly on the HEA films. This provided strong optical coupling between QDs and the films.

The exemplary results of the experiments are presented in **Figures 12–14**. The three-component (Au-Ag-Cu) film on the aluminum substrate amplified the PL of the perovskite QDs 16.5 times (**Figure 12**). A similar film on copper amplified PL 6.2 times (**Figure 13**). The six-component (Au-Ag-Cu-Al-Ti-W) HEA film on aluminum also amplified PL 8.5 times. All the films were deposited on hot substrates with a magnetic field. No significant amplification of PL was observed, for instance, for the three-component films on other substrates (Ni or Fe) or

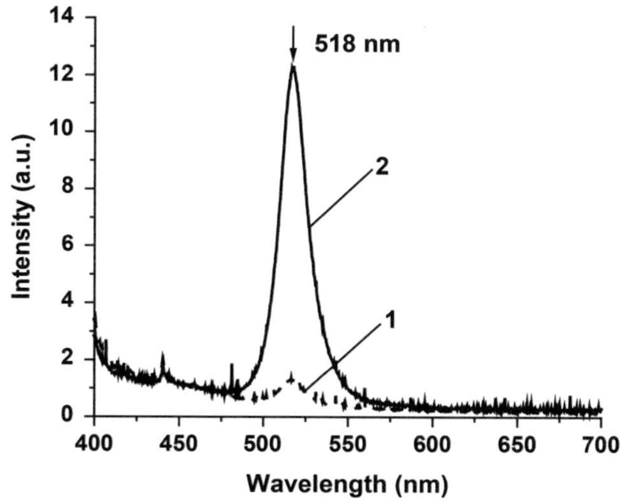

Figure 12.
The spectra of the photoluminescence (PL) of perovskite $CsPbBr_3$ pumped by a 372-nm UV laser. Curve 1 corresponds to the perovskite film spin-cast on a glass slide mounted on the top of the three-component (Au-Ag-Cu on Al substrate) HEA film. Curve 2: The perovskite film is spin-cast directly from the HEA film. PL intensity in the latter case is 16.5 times greater than in the former case.

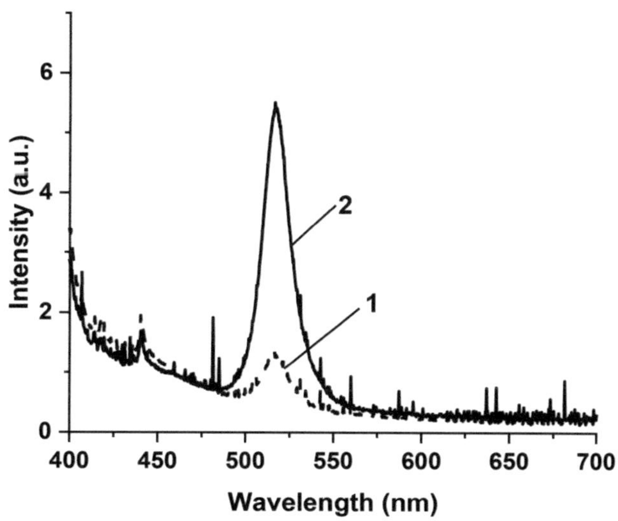

Figure 13.
The spectra of the photoluminescence (PL) of perovskite $CsPbBr_3$ pumped by a 372 nm UV laser. Curve 1 corresponds to the perovskite film spin-cast on a glass slide mounted on the top of the three-component (Au-Ag-Cu on Cu substrate) HEA film. Curve 2: The perovskite film is spin-cast directly from the HEA film. PL intensity in the latter case is 6.2 times greater than in the former case.

other six-component HEA films. As the previous analysis indicated, the prepared films were the aggregates of single-metal and double-metal alloy NPs (**Figure** 7). The amplification can be assumed to originate from the interaction between the excited QDs and the metal NPs in their vicinity. Further considerations support this assumption.

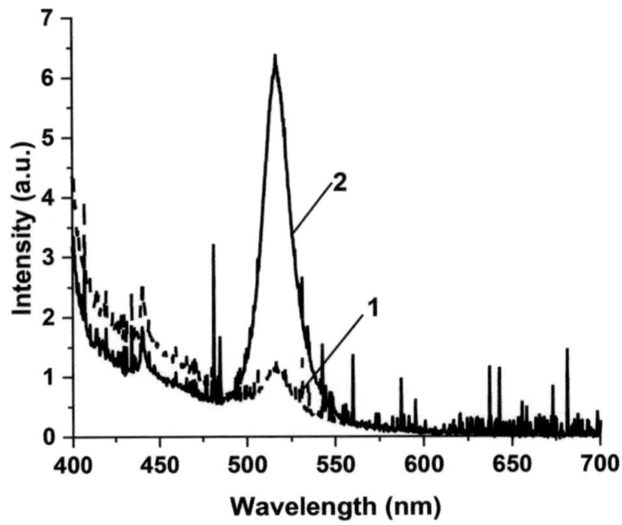

Figure 14.
The spectra of the photoluminescence (PL) of perovskite $CsPbBr_3$ pumped by a 372 nm UV laser. Curve 1 corresponds to the perovskite film spin-cast on a glass slide mounted on the top of the six-component (Au-Ag-Cu-Al-Ti-W on Al substrate) HEA film. Curve 2: The perovskite film is spin-cast directly from the HEA film. PL intensity in the latter case is 8.5 times greater than in the former case.

The polarizability χ_m at an optical frequency ω of a spherical metal NP with negative permittivity ε_m embedded in a surrounding medium with positive permittivity ε_s is defined as [29].

$$\chi_m(\omega) = 4\pi^3\left[\varepsilon_m(\omega) - \varepsilon_s(\omega)\right] / [\varepsilon_m(\omega) + 2\varepsilon_s(\omega]. \quad (1)$$

The polarizability χ_m would satisfy the LSPP resonance condition when the denominator reaches its minimum value. The imaginary part $Im\left[\varepsilon_m(\omega)\right]$ of most noble and transitional metals is small and has slowly varying values with respect to the real part $\mathrm{Re}\left[\varepsilon_m(\omega)\right]$. Therefore, LSPP resonance takes place when

$$\mathrm{Re}\left[\varepsilon_m(\omega)\right] = -2\varepsilon_s(\omega). \quad (2)$$

If the permittivity of the metal satisfies the resonance condition at the frequency of the incident UV radiation ω_{UV}, then Eqs. (1) and (2) indicate that the incident electric field is enhanced in the vicinity of the NP and the nearby QD [27, 29]. In this case, the NP acts as a nanoantenna, capturing the incident UV radiation and transferring its energy to efficiently excite the QD. On the other side, if the resonance condition in Eq. (2) takes place at the PL frequency ω_{PL} of the QD, efficient coupling between the QD and the NP occurs, leading to faster decay of the excited QD state and intense PL (Purcell effect) [27]. In our case, the perovskite QDs deposited on top of the metal films are close to the NPs of various metals. Among them, silver NPs have the SPP resonance at ω_{UV} (in the UV region 350 to 450 nm) and gold NPs at ω_{PL} (between 500 and 570 nm) [29, 30]. The conditions for the two-way PL amplification of the perovskite QDs could thus be met in both ways.

7. Conclusions

High-power Q-switched pulsed lasers were successfully used to implement the concurrent multi-beam multi-target pulsed laser deposition (CMBMT-PLD) of the high-entropy alloy (HEA) films. A six-beam system for such a process was designed and built. Additional features were introduced in the system, such as permanent magnets under the PLD targets and the substrate and a substrate heater. The magnets narrowed down the plasma plumes from the targets ablated by the laser beams and increased the material deposition rate. Three- (Au-Ag-Cu), four- (previous three + Al), five- (previous four + Ti), and six-metal (previous five + W) films were deposited on Cu, Al, Fe, and Ni substrates. Scanning electron microscopy and X-ray analysis indicated that the films were the aggregates of crystalline nanoparticles (NPs) of mostly individual metals and double-metal alloys. Heating the substrates to up to 300°C resulted in the formation of a wider variety of double-alloy nanoparticles due to higher mobility and reactivity of the deposited atomic species. The ability of the produced films to amplify photoluminescence (PL) was tested using the quantum dots (QDs) of lead-halide perovskite $CsPbBr_3$. The PL of the QDs directly deposited on the some of the HEA films and excited by a UV laser was amplified by a factor of 6 to 16. This was because of two simultaneous effects. The first was the enhancement of the incident UV radiation by the localized surface plasmon polariton resonance in the metal NPs surrounding the perovskite QD. The second was the fast decay of the excited state of the QD due to a resonant coupling between the QD and the NPs in its vicinity. The findings described in this chapter might be useful for applications in optical sensing.

Acknowledgements

A. Darwish and S. Sarkisov appreciate financial support from the US Air Force Office of Scientific Research Grant No. FA9550-23-1-0611, Army Research Office Grant No. W911NF-24-1-0253, partial support from Dillard University Minority Health and Health Disparity Research Center MHHDRC, and DOE INSIGHT Center. D. Patel's research was sponsored by the NSF Grant No. 2200426. The views and conclusions contained in this document are those of the authors and should not be interpreted as representing the official policies, either expressed or implied, of the U.S. Government. The U.S. government is authorized to reproduce and distribute reprints for government purposes, notwithstanding any copyright notation herein.

Conflict of interest

The authors declare no conflict of interest.

Author details

Abdalla M. Darwish[1], Sergey S. Sarkisov[2]* and Darayas N. Patel[3]

1 Physics and Pre-Engineering Department, Dillard University, New Orleans, LA, USA

2 SSS Optical Technologies, LLC, Huntsville, AL, USA

3 Math and Computer Science Department, Oakwood University, Huntsville, AL, USA

*Address all correspondence to: mazillo@hotmail.com

References

[1] Shepelin NA, Tehrani ZP, Ohannessian N, Schneider CW, Pergolesi D, Lippert T. A practical guide to pulsed laser deposition. Chemical Society Reviews. 2023;**52**:2294. DOI: 10.1039/d2cs00938b

[2] Korvus Technology. Pulsed Laser Deposition Process of Thin Films: An Overview [Internet]. High Wycombe, United Kingdom: Korvus Technology Ltd; 2022. Available from: https://korvustech.com/pulsed-laser-deposition-process/ [Accessed: June 3, 2025]

[3] Darwish A, Sarkisov S, Mele P. Nano-Composite Thermo-Electric Energy Converter and Fabrication Method Thereof. US Patent No 10283691 B2; 2019. Washington, DC, USA: USPTO; 2019. Available from: https://patents.google.com/patent/US10283691B2/en?oq=10283691 [Accessed: June 3, 2025]

[4] Darwish A, Sarkisov S. Method and Apparatus for Open-Air Pulsed Laser Deposition. US Patent No 10316403 B2; 2019. Washington, DC, USA: USPTO; 2019. Available from: https://patents.google.com/patent/US10316403B2/en?oq=10316403 [Accessed: June 3, 2025]

[5] Darwish A, Sarkisov S. Multiple Beam Pulsed Laser Deposition of Composite Films. US Patent No 11059128 B2; 2021. Washington, DC, USA: USPTO; 2019. Available from: https://patents.google.com/patent/US11059128B2/en?oq=11059128 [Accessed: June 3, 2025]

[6] Nikov RG, Dikovska AO, Avdeev GV, Atanasova GB, Karashanova DB, Amoruso S, et al. Single-step fabrication of oriented composite nanowires by pulsed laser deposition in magnetic field. Materials Today Communications. 2021;**26**:101717. DOI: 10.1016/j.mtcomm.2020.101717

[7] Huang G, Lu Y, Wang S, Ding F, Wei S, Mi C, et al. Pulsed laser deposition of high-quality diamond-like carbon films under an inhomogeneous magnetic field. Journal of Physics Conference Series. 2021;**1775**(1):012010. DOI: 10.1088/1742-6596/1775/1/012010

[8] Tao S, Dai H, Song M, Liu H, Wei X. Magnetic and magneto-optical properties of CeYIG thin films grown by pulsed laser deposition on quartz substrates. Journal of Superconductivity and Novel Magnetism. 2021;**30**(7):1915-1919. DOI: 10.1007/s10948-017-3993-9

[9] Miracle DB. High entropy alloys as a bold step forward in alloy development. Nature Communications. 2019;**10**:1805. DOI: 10.1038/s41467-019-09700-1

[10] Chen W-H, Chen Y-S, Wu P-J, Chen C-H, Shih C-F, Yang C-S. 3D printed high-entropy plasmonic structures for cutting-edge 6G modulation components on silicon wafers via electromagnetically induced reflection. In: Proceedings of SPIE Conference on Terahertz, RF, Millimeter, and Submillimeter-Wave Technology and Applications XVII; 11 March 2024; San Francisco, USA. Bellingham, WA: Proc. SPIE 12885; 2024. p. 288513. DOI: 10.1117/12.2691576

[11] Calzolari A, Oses C, Toher C, Esters M, Campilongo X, Stepanoff SP, et al. Plasmonic high-entropy carbides. Nature Communications. 2022;**13**:5993. DOI: 10.1038/s41467-022-33497-1

[12] Lu C-Y, Sung Y-C, Yeh C-H, Chen Y-H, Kuan C-Y, Hung K-T, et al. Terahertz characterization of nano-scale high-entropy alloy films and their

high-contrast grating applications. AIP Advances. 2024;**14**:035036. DOI: 10.1063/5.0188695

[13] Cui K, Zhang Y. High-entropy alloy films. Coatings. 2023;**13**(3):635. DOI: 10.3390/coatings13030635

[14] Touaibia DE, Bouissil A, Achache S, Sanchette F, El Garah M. High-entropy films by magnetron sputtering: Deposition parameters, oxidation resistance enhancement, and applications. In: Thin Films - Fundamentals, Deposition Techniques and Applications. London, UK: IntechOpen; 2024. DOI: 10.5772/intechopen.1007963

[15] Gunina EV, Larin AO, Kustov PN, Nomine A, Ghanbaja J, Bruyere S, et al. Fabrication of resonant highentropy alloy nanospheres by the laser ablation technique. In: 2022 Sixteenth International Congress on Artificial Materials for Novel Wave Phenomena (Metamaterials); 12-17 September 2022; Siena, Italy. New York, USA: IEEE; 2022. DOI: 10.1109/Metamaterials54993.2022.9920910

[16] Kurt J. Lesker Company. Sputtering Targets [Internet]. Pittsburg, PA, USA: Kurt J. Lesker Company; 2025. Available from: https://www.lesker.com/materials-division.cfm?section=sputtering-targets [Accessed: June 5, 2025]

[17] Wang S, Xu Y, Chen R, Zhu M, Wang M, Cao M, et al. Highly efficient and stable Eu^{3+}-doped $CsPbBr_3$/Cs_4PbBr_6 perovskites for white light-emitting diodes. Coatings. 2022;**12**:512. DOI: 10.3390/coatings12040512

[18] dela Cruz-Pina E, Pareja-Rivera C, Zugasti-Fernandez D, Rodriguez-Gomez A, Solis-Ibarra D, Reyes-Esqueda JA. Photoluminescence enhancement of quasi-2D perovskite films by plasmonic silver nanoparticles. Journal of Materials Chemistry C. 2025;**13**(15):1-9. DOI: 10.1039/D4TC04288C

[19] Wang Z, Dong Z, Gu Y, Chang Y-H, Zhang L, Li L-J, et al. Giant photoluminescence enhancement in tungsten-diselenide–gold plasmonic hybrid structures. Nature Communications. 2016;**7**:11283. DOI: 10.1038/ncomms11283

[20] Lee GY, Jung K, Ho Jang HS, Kyhm J, Han IK, Park B, et al. Up conversion luminescence enhancement in plasmonic architecture with random assembly of metal nanodomes. Nanoscale. 2016;**8**:2071-2080. DOI: 10.1039/C5NR05628D

[21] Łapiński M, Czubek J, Drozdowska K, Synak A, Sadowski W, Kościelska B. Plasmon-enhanced photoluminescence from TiO_2 and TeO_2 thin films doped by Eu^{3+} for optoelectronic applications. Beilstein Journal of Nanotechnology. 2021;**22**(12):1271-1278. DOI: 10.3762/bjnano.12.94

[22] Liu J, Wang Q, Sang X, Hu H, Li S, Zhang D, et al. Modulated luminescence of lanthanide materials by local surface plasmon resonance effect. Nanomaterials. 2021;**11**(4):1037. DOI: 10.3390/nano11041037

[23] Semeniak D, Cruz DF, Chilkoti A, Mikkelsen MH. Plasmonic fluorescence enhancement in diagnostics for clinical tests at point-of-care: A review of recent technologies. Advanced Materials. 2023;**35**:2107986. DOI: 10.1002/adma.202107986

[24] Cabello-Olmo E, Higashino M, Murai S, Tanaka K, Lozano G, Míguez H. Collective plasmonic resonances enhance the photoluminescence of rare-earth

nanocrystal films processed by ultrafast annealing. Chemical Communications. 2023;**59**:1289-1292. DOI: 10.1039/D2CC04779A

[25] Zhang L, Liu J, Tian L, Zhang D, Wang Q. Plasmon enhanced luminescence of Tb/Eu co-doped film by Au NRs-PVA nanocomposite film. Biomedical Optics Express. 2023;**15**(1):306-318. DOI: 10.1364/BOE.512053

[26] Piotrowski P, Buza M, Nowaczyński R, Kongsuwan N, Surma HB, Osewski P, et al. Ultrafast photoluminescence and multiscale light amplification in nanoplasmonic cavity glass. Nature Communications. 2024;**15**:3309. DOI: 10.1038/s41467-024-47539-3

[27] Montaño-Priede JL, Zapata-Herrera M, Esteban R, Zabala N, Aizpurua J. An overview on plasmon-enhanced photoluminescence via metallic nanoantennas. Nano. 2024;**13**(26):4771-4794. DOI: 10.1515/nanoph-2024-0463

[28] Ahmed WW, Cao H, Xu C, Farhat M, Amin M, Li X, et al. Machine learning assisted plasmonic metascreen for enhanced broadband absorption in ultra-thin silicon films. Light: Science & Applications. 2025;**14**:42. DOI: 10.1038/s41377-024-01723-8

[29] Plasmonic LY, Optics. Theory and Applications. Billingham, WA, USA: SPIE Press; 2017. 250 p. DOI: 10.1117/3.2263757

[30] Williams EK, Ila D, Sarkisov S, Curley M, Cochrane JC, Poker DB, et al. Study of the effects of MeV Ag, Cu, Au, and Sn implantation on the optical properties of $LiNbO_3$. Nuclear Instruments and Methods in Physics Research B. 1998;**141**:268-273. DOI: 10.1016/S0168-583X(98)00150-5